Applied Biomechanics and Sports Sciences

Applied Biomechanics and Sports Sciences

Guest Editors

Filipe Conceição
Ricardo J. Fernandes
Pedro Jimenez-Reyes

MDPI

Basel • Beijing • Wuhan • Barcelona • Belgrade • Novi Sad • Cluj • Manchester

Guest Editors
Filipe Conceição
Faculty of Sport
University of Porto
Porto
Portugal

Ricardo J. Fernandes
Faculty of Sport
University of Porto
Porto
Portugal

Pedro Jimenez-Reyes
Sport Sciences Research
Centre
Rey Juan Carlos University
Madrid
Spain

Editorial Office
MDPI AG
Grosspeteranlage 5
4052 Basel, Switzerland

This is a reprint of the Special Issue, published open access by the journal *Applied Sciences* (ISSN 2076-3417), freely accessible at: https://www.mdpi.com/journal/applsci/special_issues/89F83MQCD6.

For citation purposes, cite each article independently as indicated on the article page online and as indicated below:

Lastname, A.A.; Lastname, B.B. Article Title. *Journal Name* **Year**, *Volume Number*, Page Range.

ISBN 978-3-7258-6756-1 (Hbk)
ISBN 978-3-7258-6757-8 (PDF)
https://doi.org/10.3390/books978-3-7258-6757-8

Contents

About the Editors

Filipe Conceição

Filipe Conceição holds a PhD in Sport Sciences and is a Lecturer at the Faculty of Sport, University of Porto (FADEUP), Portugal, where he also serves as Head of the Athletics Department. He is a member of the Biomechanics Laboratory (LABIOMEP), where his research focuses on experimental and computational biomechanics, strength and power development, and the neuromechanical determinants of sprinting and jumping performance. His work combines laboratory-based analysis with applied approaches to optimize training, monitor performance, and understand the musculoskeletal mechanisms underlying high-intensity movement. In addition to his academic career, he coaches elite Portuguese sprinters and jumpers and provides consultancy in strength and conditioning to high-level athletes and professional teams worldwide. He has supervised several master's and doctoral projects in sports performance and collaborates with international research groups in biomechanics and applied sports sciences.

Ricardo J. Fernandes

Ricardo J. Fernandes is an Associate Professor in the Sports Training area at the Faculty of Sport, University of Porto, Portugal and member of the Biomechanics Laboratory (LABIOMEP). His research interests encompass the biomechanics, physiology and energetics of cyclic sports, with a particular focus on efficiency, technique and performance modeling. He has authored numerous scientific publications in applied sport sciences and actively participates in international collaborations and editorial duties. His work bridges experimental research and real-world training applications, contributing to the development of innovative methodologies for performance monitoring and optimization. Throughout his career, he has supervised a large number of graduate and postgraduate theses and has played a major role in advancing the understanding of aquatic and terrestrial physiology, biomechanics and training science at both the academic and applied levels.

Pedro Jiménez-Reyes

Pedro Jiménez-Reyes is an Associate Professor of Sports Training Methodology at Rey Juan Carlos University in Madrid, Spain. He holds a PhD with a European Mention in Sports Sciences and a Master's in High-Performance Sports and Physical and Sports Performance. His research focuses on the biomechanics and physiology of human movement, with particular emphasis on force–velocity profiling in sprinting and jumping, and the individualization of strength training based on these profiles. Over the past 15 years, he has developed performance-driven solutions for elite athletes, mainly in track and field, through close collaboration with coaches and performance scientists. He has undertaken more than 25 months of research stays at leading international sports performance centers and has authored over 100 peer-reviewed publications and several book chapters. In addition to his academic role, he leads a biomechanics and sprint-training group in Madrid and provides consultancy in physical fitness assessment, load monitoring, and strength training for elite athletes and teams worldwide.

Preface

The Reprint "Applied Biomechanics and Sports Sciences" was conceived to integrate multidisciplinary perspectives on the mechanisms underlying human performance. The contributing authors explore how biomechanical and neuromuscular principles can translated into practical strategies to optimize movement efficiency, training prescription and injury prevention. This collection brings together experimental and applied studies that span a diversity of sports and methodological approaches, illustrating how advances in biomechanics and technology can inform evidence-based coaching and performance enhancement. As Guess Editors, we aimed to provide a scientific platform that bridges laboratory-based research and real-word applications, highlighting the growing relevance of biomechanics as a cornerstone of modern sports science. We would like to express our gratitude to all the researchers, reviewers and editorial staff who have contributed to making this Reprint possible.

Filipe Conceição, Ricardo J. Fernandes, and Pedro Jimenez-Reyes
Guest Editors

applied sciences

MDPI

Editorial

Applied Biomechanics and Sports Sciences: Closing Editorial

Filipe Conceição [1,2], Ricardo J. Fernandes [1,2,*] and Pedro Jimenez-Reyes [3]

1 Centre of Research, Education, Innovation and Intervention in Sport, CIFI2D, Faculty of Sport, University of Porto, 4200-450 Porto, Portugal; filipe@fade.up.pt
2 Porto Biomechanics Laboratory, LABIOMEP-UP, Faculty of Sport, University of Porto, 4200-450 Porto, Portugal
3 Center for Sport Studies, Rey Juan Carlos University, 28933 Madrid, Spain
* Correspondence: ricfer@fade.up.pt

1. Introduction

The integration of biomechanics and sports sciences has accelerated markedly in recent years largely due to advances in sensor technology, data analytics, and computational modeling. Wearable devices, such as inertial measurement units and portable electromyography systems, have enhanced the capacity to capture real-time data in ecologically valid conditions, supporting both training and rehabilitation practices [1,2]. Parallel to this, artificial intelligence and machine learning approaches are increasingly employed in performance prediction and individualized training design [3,4]. However, substantial gaps remain since translating complex biomechanical findings into practical tools that coaches can apply with confidence continues to be a challenge, while longitudinal multi-level studies are still underrepresented [5]. Furthermore, there is limited integration across biomechanics, physiology and psychology, restricting our ability to holistically model athlete development. This Special Issue sought to highlight research directions that combine methodological rigor with real-world impact, in line with contributions to applied biomechanics, aquatic performance, and sprint profiling.

2. An Overview of Thematic Contributions

The contributions to this Special Issue can be grouped into four thematic clusters that mirror emergent research priorities. The first relates to accessible monitoring technologies, exemplifying the push toward accurate but low-cost tools that broaden the application of biomechanics beyond laboratory settings [6]. This perspective is consistent with our work on affordable and reliable measurement solutions for neuromuscular assessment in applied contexts [7,8]. A second cluster focuses on predictive modeling and longitudinal profiling, a field where advanced statistics and computational models are increasingly applied to monitor training adaptations and forecast record progression [9,10]. This resonates with our extensive contributions to swimming physiology and energetics, where integrative models have been developed to link biomechanics and bioenergetics to performance outcomes [11].

A third theme highlights underrepresented populations and contexts, such as female athletes, youth, and those recovering from injury. Addressing these gaps is essential to closing the persistent imbalance in sex- and age-based data in sport science [12]. The attention to such groups in this Special Issue also reflects the editorial vision of creating inclusive knowledge that informs talent identification and equitable training strategies. Finally, several studies focus on biomechanical variability and asymmetry, with applications in explosive actions and rehabilitation. This line of research reflects the necessity of accounting for functional imbalances when designing interventions [13,14]. Our research on sprint

 https://doi.org/10.3390/app151810059

force–velocity profiling also exemplifies this principle, highlighting how individualized mechanical profiles can guide strength and speed training, optimizing performance while reducing risk of injury [8,15].

3. Conclusions and Future Directions

The contributions in this Special Issue illustrate the breadth of contemporary applied biomechanics, spanning validation of new tools, predictive analytics, profiling of youth athletes and clinical rehabilitation. A recurrent theme is that biomechanics must not be considered in isolation, but should be integrated with physiology, psychology, and training methodology to capture the complexity of sport performance. Persistent challenges include the need for standardized approaches across laboratory and field contexts, the greater inclusion of underrepresented populations and the translation of sophisticated methods into user-friendly formats for practitioners [5,16]. Future research should prioritize longitudinal and integrative frameworks that track adaptation over time, advanced computational tools (including artificial intelligence and force–velocity profiling for individualized diagnostics; [17]) and equitable approaches tailored to women, youth, and clinical populations [12]. In addition, scalable wearable technologies must be designed with attention to ethical and practical factors, such as affordability and data privacy [2,18]. Building on the expertise of this Special Issue's Guest Editors, the future of applied biomechanics should continue to move toward integrated, accessible and impactful applications that benefit both elite sport and broader human performance contexts.

Conflicts of Interest: The authors declare no conflict of interest.

List of Contributions:

1. Abelleira-Lamela, T.; Marcos-Pardo, P.J.; Abraldes, J.A.; González-Gálvez, N.; Espeso-García, A.; Esparza-Ros, F.; Vaquero-Cristóbal, R. Comparative Electromyographic Analysis in Leg Press of Traditional Fitness Equipment, Traditional Outdoor Fitness Equipment, and a New Model of Outdoor Fitness Equipment in Trained Young Men. *Appl. Sci.* **2024**, *14*, 7390.
2. Costa, M.J.; Quinta-Nova, L.; Ferreira, S.; Costa, A.M.; Santos, C.C. Trend Forecasting in Swimming World Records and in the Age of World Record Holders. *Appl. Sci.* **2024**, *14*, 9492.
3. Castaño-Zambudio, A.; Repullo, C.; Jiménez-Reyes, P. Enhancing Acceleration Capabilities in Professional Women's Football Players: A Comparative Analysis of Game-Based Versus Resisted Sprint Trainings. *Appl. Sci.* **2024**, *14*, 10327.
4. Ramasamy, Y.; Wei, Y.M.; Towler, H.; King, M. Intra-Individual Variation in the Jump Smash for Elite Malaysian Male Badminton Players. *Appl. Sci.* **2025**, *15*, 844.
5. Delgado-García, G.; Pérez-Castilla, A.; Rojas-Ruiz, F.J.; Navarro-Marchal, I.; Caballero-Villalta, A. Validity of a User-Friendly Spreadsheet Designed for an In-Depth Analysis of Countermovement Bipodal Jumps. *Appl. Sci.* **2025**, *15*, 1519.
6. Silva, M.; Antunes, H.D.; Sousa, A.; Nakamura, F.Y.; Sampaio, A.R.; Pimenta, R. Profiling of Physical Qualities of Highly Trained Portuguese Youth Soccer Players. *Appl. Sci.* **2025**, *15*, 5414.
7. Romero Padron, M.A.; Jorgensen, A.; Werner, D.M.; Tao, M.A.; Wellsandt, E. Knee Loading Asymmetries During Descent and Ascent Phases of Squatting After ACL Reconstruction. *Appl. Sci.* **2025**, *15*, 7780.
8. Noronha, F.; Canossa, S.; Carvalho, D.D.; Monteiro, A.S.; Afonso, J.; Castro, F.; Fernandes, R. Sex and Age Disparities in Water Polo-Related Skills. *Appl. Sci.* **2025**, *15*, 9381.

References

1. Jaén-Carrillo, D.; Pérez-Castilla, A.; García-Pinillos, F. Wearable and Portable Devices in Sport Biomechanics and Training Science. *Sensors* **2024**, *24*, 4616. [CrossRef] [PubMed]
2. Van Hooren, B.; Goudsmit, J.; Restrepo, J.; Vos, S. Real-time feedback by wearables in running: Current approaches, challenges and suggestions for improvements. *J. Sports Sci.* **2020**, *38*, 214–230. [CrossRef] [PubMed]

3. Claudino, J.G.; Capanema, D.; de Souza, T.V.; Serrão, J.C.; Machado Pereira, A.C.; Nassis, G.P.; Filho, C.A.C. Current approaches to the use of artificial intelligence for injury risk assessment and performance prediction in team sports: A systematic review. *Sports Med.–Open* **2019**, *5*. [CrossRef]
4. Jianjun, Q.; Isleem, H.F.; Almoghayer, W.J.K.; Khishe, M. Predictive athlete performance modeling with machine learning and biometric data integration. *Sci. Rep.* **2025**, *15*, 16365. [CrossRef] [PubMed]
5. Bishop, D.; Burnett, A.; Farrow, D.; Gabbett, T.; Newton, R. Sports science research: Methodological issues and applications. *Sports Med.* **2019**, *49*, 1–16. [CrossRef]
6. Camomilla, V.; Bergamini, E.; Fantozzi, S.; Vannozzi, G. Trends supporting the in-field use of wearable inertial sensors for sport performance evaluation: A systematic review. *Sensors* **2018**, *18*, 873. [CrossRef] [PubMed]
7. Balsalobre-Fernández, C.; Glaister, M.; Lockey, R.A. The validity and reliability of an iPhone app for measuring vertical jump performance. *J. Sports Sci.* **2016**, *33*, 1574–1579. [CrossRef] [PubMed]
8. Silva, M.P.; Fonseca, P.; Fernandes, R.J.; Conceição, F. Is running technique important to mitigate hamstring injuries in football players? *Appl. Sci.* **2024**, *14*, 11643. [CrossRef]
9. Allen, S.V.; Vandenbogaerde, T.J.; Hopkins, W.G. Career performance trajectories of Olympic swimmers: Benchmarks for talent development. *Eur. J. Sport Sci.* **2014**, *14*, 643–651. [CrossRef]
10. Romann, M.; Müller, S. Longitudinal performance trajectories of young female 60 m sprinters using linear mixed effects models. *Front. Sports Act. Living* **2024**, *6*. [CrossRef]
11. Fernandes, R.J.; Carvalho, D.D.; Figueiredo, P. Training zones in competitive swimming: A biophysical approach. *Front. Sports Act. Living* **2024**, *6*, 1363730. [CrossRef] [PubMed]
12. Cowley, E.S.; Olenick, A.A.; McNulty, K.L.; Ross, E.Z. "Invisible sportswomen": The sex data gap in sport and exercise science research. *Women Sport Phys. Act. J.* **2021**, *29*, 146–151. [CrossRef]
13. Morin, J.-B.; Samozino, P. Interpreting power-force-velocity profiles for individualized and specific training. *Int. J. Sports Physiol. Perform.* **2016**, *11*, 267–272. [CrossRef] [PubMed]
14. Mendiguchia, J.; Castaño-Zambudio, A.; Jiménez-Reyes, P.; Morin, J.-B.; Edouard, P.; Conceição, F.; Tawiah-Dodoo, J.; Colyer, S.L. Can We Modify Maximal Speed Running Posture? Implications for Performance and Hamstring Injury Management. *Int. J. Sports Physiol. Perform.* **2022**, *17*, 374–383. [CrossRef] [PubMed]
15. Jiménez-Reyes, P.; Garcia-Ramos, A.; Párraga-Montilla, J.A.; Morcillo-Losa, J.A.; Cuadrado-Peñafiel, V.; Castaño-Zambudio, A.; Samozino, P.; Morin, J.-B. Seasonal changes in the sprint acceleration force-velocity profile of elite male soccer players. *J. Strength Cond. Res.* **2022**, *36*, 70–74. [CrossRef] [PubMed]
16. Windt, J.; Gabbett, T.J. How do training and competition workloads relate to injury? The workload–injury aetiology model. *Br. J. Sports Med.* **2017**, *51*, 428–435. [CrossRef] [PubMed]
17. Rojas-Valverde, D.; Gutiérrez-Vargas, R.; Sánchez-Ureña, B.; Crowe, J.; Gutiérrez-Vargas, J.C. Application of machine learning in sports: A bibliometric and content analysis. *Appl. Sci.* **2022**, *12*, 3483. [CrossRef]
18. Halson, S.L.; Mujika, I.; Meeusen, R. Monitoring training load: The past, the present, and the future. *Int. J. Sports Physiol. Perform.* **2019**, *14*, 2–11. [CrossRef]

applied sciences

MDPI

Article

Sex and Age Disparities in Water Polo-Related Skills

Francisco Noronha [1,2], Sofia Canossa [1], Diogo D. Carvalho [1,2], Ana Sofia Monteiro [1,2], José Afonso [1], Flávio Castro [3] and Ricardo Fernandes [1,2,*]

1 Centre of Research, Education, Innovation and Intervention in Sport, Faculty of Sport, University of Porto, 4099-002 Porto, Portugal; up201302164@fade.up.pt (F.N.); soficanossa@gmail.com (S.C.); diogoac@fade.up.pt (D.D.C.); up201406202@fade.up.pt (A.S.M.); jneves@fade.up.pt (J.A.)
2 Porto Biomechanics Laboratory, Faculty of Sport, University of Porto, 4099-002 Porto, Portugal
3 Aquatic Sports Research Group, School of Physical Education, Physiotherapy and Dance, Universidade Federal do Rio Grande do Sul, Porto Alegre 90010-150, Brazil; souza.castro@ufrgs.br
* Correspondence: ricfer@fade.up.pt

Abstract

Existing research points to sex-specific biological and physiological differences, but the effects of sex and age on water polo-specific motor skills have not yet been sufficiently investigated. The current study investigated sex- and age-related disparities in in-water vertical jump, change of direction ability, sprint performance, aerobic fitness and shot velocity, hypothesising that these variables are related to the players' specific motor performance. Sixty-six players (47 males, 19 females) were split into an adult and a young group (each with n = 33). Two-way (sex and age as factors) analysis of variance and Tukey's post hoc test were employed to evaluate the influence of sex and age on specific water polo skill performance. The male players' in-water vertical jump (135.4 ± 8.1 vs. 121.5 ± 6.1 cm), shot velocity with previous displacement (68.3 ± 6.4 vs. 51.0 ± 2.5 km·h^{-1}) and without displacement (68.5 ± 5.7 vs. 52.2 ± 2.9 km·h^{-1}), 10 m sprint (5.6 ± 0.9 vs. 6.6 ± 0.4 s), aerobic fitness (311.7 ± 92.1 vs. 235.7 ± 70.3 m) and change of direction (3.3 ± 0.4 vs. 3.9 ± 0.4 s) scores were superior to those of female players. No significant age-related differences were observed in shot velocity with previous displacement ($F(1,53) = 2.124$, $p = 0.151$, $\eta^2p = 0.039$) or 10 m sprint performances ($F(1,62) = 0.935$, $p = 0.337$, $\eta^2p = 0.015$). Male players outperformed females in most water polo-specific motor skills, while age-related differences were limited.

Keywords: physiology; agility; skills; endurance

1. Introduction

Research on sex and age disparities in sports spans multiple disciplines and examines their impact on athletic performance. However, persistent participation gaps suggest underlying patterns that remain poorly understood [1]. Available data indicate that equalising performance between the sexes in sport is challenging, as adult men generally perform 10–30% better than trained women across various sports [2]. Traditional discussions often focus on biological sex due to anatomical and physiological differences caused by sex chromosomes and hormones [2,3]. These discrepancies emerge during puberty, when hormonal changes shape physical development and influence the performance trajectories of male and female players throughout ageing [4,5]. While biological sex establishes fundamental physical distinctions, the influence of age introduces additional complexity to performance development and athletic trajectories.

Age plays a crucial role in athletic performance, influencing physical and sport-specific motor skills, as well as career longevity. Peak performance is typically achieved in early adulthood, although this varies according to the specific demands of each sport [6]. These differences highlight the need for age-specific training programmes to optimise performance potential [7]. In water polo, technical skills such as ball handling, passing, shooting, eggbeater kicking, and defensive and offensive actions depend on physiological, anatomical and biomechanical factors [8,9]. Consequently, variations in technical skills across developmental stages impact training strategies, competition levels and talent identification. These factors, shaped by both sex and age, influence performance and athlete development. Moreover, men and women may exhibit different performance profiles even within the same age group, due to differences in maturation [10] and the ageing process, which affects the sexes differently [11].

A closer examination of the data may reveal a more complex and potentially unexpected pattern that challenges conventional assumptions. In water polo, the importance of specific technical skills in game situations is well established, as they are essential to overall performance and player success [12,13]. However, a review of the literature indicates that sex- and age-related disparities in these skills are not yet fully understood [14]. This study conducted a comprehensive investigation into such disparities in the technical abilities of water polo players. It also explored the physiological and experiential factors underlying these differences, as well as their impact on performance and developmental trajectories. By addressing these gaps, the study provides new insights into how skill acquisition and refinement may differ based on age and sex. These findings highlight the need to examine sex- and age-specific effects on performance to inform better training strategies.

Understanding how sex and age influence sport-specific technical skills is essential for optimising training, performance and talent development in elite water polo [15]. While the existing literature emphasises physiological differences between the sexes and age-related performance trends, few studies have systematically investigated how these factors affect key technical skills in water polo [16]. As a sport that uniquely combines strength, agility, endurance and tactical execution in an aquatic environment, it is crucial to assess performance using specific metrics such as in-water vertical jump, swimming velocity, shot velocity, and change of direction [17]. Furthermore, female and younger players remain underrepresented in performance research, resulting in a lack of evidence-based approaches tailored to their needs [18]. To address these research gaps and better understand how sex and age jointly shape technical performance in water polo, this study adopted a targeted empirical approach.

Previous research on athletic performance, acute exercise, and training in water polo has focused on male players or neglected sex-related differences altogether. As a result, substantial gaps remain in our understanding of female players physiological characteristics, performance thresholds, and responses to exercise and training relative to their male counterparts. Moreover, the influence of ageing on sex-specific differences in these areas requires further investigation. The aim of the present study was to determine whether sex- and age-related disparities in water polo are as pronounced as those reported in other sports. We addressed this question by analysing empirical data within a framework that not only confirms established trends but also highlights unexpected deviations. We hypothesised that sex and age influence the performance of water polo-specific technical skills and that their interaction also contributes to performance outcomes.

2. Materials and Methods

A total of 66 water polo players with experience of both national and international competitions volunteered to participate in the study. The mean (±SD) age of the males and females were 21.17 ± 1.0 and 19.11 ± 1.0 years, respectively. All players trained for at least 5 sessions and played one game per week. The study employed a cross-sectional design, incorporating both male and female water polo players from different age categories (adult 25.24 ± 1.0 and young 15.91 ± 0.1 years of age), enabling a direct comparison of performance outcomes and a robust evaluation of the proposed hypotheses concerning the main effects of sex, age and their potential interaction.

The independent variables—sex and age—were selected based on their well-documented influence on physiological development and athletic performance trajectories. Neuromuscular differences may contribute to this, but motor unit activity was not measured, so the mechanisms remain speculative. The dependent variables—specific technical skills—were chosen because they reflect fundamental capacities required for success in water polo and offer reliable quantifiable measures of sport-specific performance. To rigorously test the hypotheses, we developed and implemented a battery of water polo-specific technical skill assessments designed to capture the differences between sex and age groups. These assessments comprised in-water vertical jump, change of direction ability, 10 m sprint, aerobic fitness (measured via a field-based protocol) and shot velocity, all of which are recognised as key performance indicators in high-level water polo competition. Familiarisation sessions were not deemed necessary, as all participants had extensive experience with the tested skills, which are regularly integrated into both training sessions and competitive play.

2.1. Experimental Design

Players began the testing session with a standardised 500 m low-to-moderate intensity warm-up, performed in a 25 m indoor swimming pool maintained at 27 °C with a uniform depth of 1.90 m. Immediately following the warm-up, participants completed six water polo-specific technical skill tests (Figure 1), administered in a fixed order to ensure consistency across subjects. These were (i) in-water vertical jump height, assessed using a Sony Handycam HDR-PJ530, Tokyo, Japan digital video camera (1920 × 1080 p resolution) positioned laterally to the testing zone. Each player performed two maximal-effort jumps from a vertical starting position, with the highest value recorded for analysis. The technical error of measurement (TEM) for this test was 0.9 cm [19]; (ii) shooting velocity with and (iii) without displacement, which were assessed in the presence of a goalkeeper to replicate game-like conditions. For each condition, players executed six maximal shots from a 5 m distance [20]. Shot velocity was measured using a 10 Hz radar (Applied Concepts, Inc., Flat Salkerpro, TX, USA), positioned 8 m from the goal, with a sensitivity of 0.045 $m{\cdot}s^{-1}$ and a calculated TEM of 2.6%; (iv) 10 m front crawl sprint performance, as measured from three maximal trials, with the fastest time retained for analysis (TEM: 0.1 s); (v) aerobic fitness, which was assessed via the multistage shuttle swim test, a progressive, intermittent protocol previously validated for aquatic sports contexts [21]; (vi) change of direction ability, which was evaluated using a water polo-specific test protocol [22], with three trials completed and the best time recorded (TEM: 0.1 s).

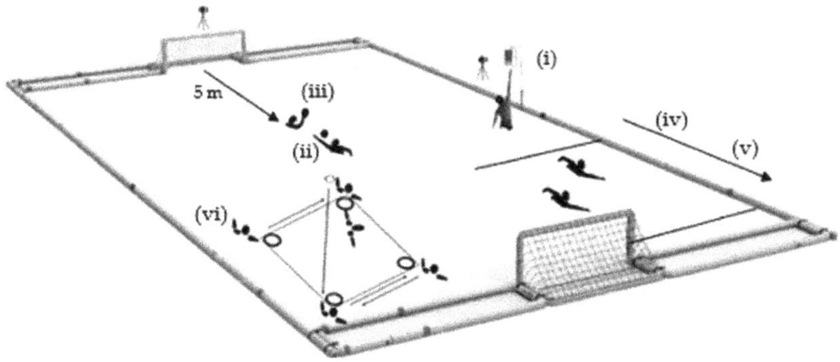

Figure 1. Representation of the experimental evaluations of specific technical skills.

2.2. Statistical Analysis

Descriptive statistics, including mean ± standard deviation and 95% confidence intervals, were calculated and reported for all performance variables to provide a comprehensive summary of central tendency and dispersion. A two-way analysis of variance (2 × 2 ANOVA) was employed to examine the main effects of sex and age group, as well as their interaction, on water polo-specific technical skill performance. When significant differences were identified, Tukey's post hoc tests were applied to determine pairwise contrasts between groups. The assumptions underlying the use of parametric tests—namely, normality of distribution and homogeneity of variances—were assessed using the Shapiro–Wilk test and visual inspection of residual plots. All statistical analyses were performed using IBM SPSS Statistics version 29.0 (IBM Corp., Armonk, NY, USA), with the threshold for statistical significance set at $p < 0.05$. Additionally, Cohen's d was computed to quantify effect sizes and enhance interpretation of group differences. Effect sizes were classified as small (<0.2), moderate (0.2–0.5), large (0.5–1.0), or very large (>1.0), following established conventions [23].

3. Results

Sex-related disparities were observed across all evaluated water polo-specific technical skills (Table 1). Male players consistently demonstrated superior performance compared to their female counterparts in each of the assessed tests. Specifically, males achieved higher scores for the in-water vertical jump ($F(1,62) = 51.297$, $p < 0.001$, $\eta^2p = 0.453$), as well as in shot velocity both with ($F(1,53) = 107.312$, $p < 0.001$, $\eta^2p = 0.669$) and without previous displacement ($F(1,52) = 134.790$, $p < 0.001$, $\eta^2p = 0.722$). Significant sex differences were also observed in 10 m sprint performance ($F(1,62) = 19.496$, $p < 0.001$, $\eta^2p = 0.239$), aerobic fitness, as measured by the shuttle swim test ($F(1,61) = 14.986$, $p < 0.001$, $\eta^2p = 0.197$), and change of direction ability ($F(1,62) = 32.953$, $p < 0.001$, $\eta^2p = 0.347$). These findings highlight the presence of robust sex-related disparities across key physical performance domains relevant to water polo. The magnitude of the observed effects, as reflected by the large partial eta squared values, suggests that sex is a substantial determinant of technical and physical performance in this population.

Table 1. Descriptive statistics (mean ± SD) of specific motor abilities performance by sex.

Variables	In-Water Vertical Jump (cm)	Shot Velocity with Previous Displacement (km·h^{-1})	Shot Velocity Without Displacement (km·h^{-1})	10 m Sprint (s)	Aerobic Fitness (m)	Change of Direction (s)
Male	135.4 ± 8.1	68.3 ± 6.4	68.5 ± 5.7	5.6 ± 0.9	311.7 ± 92.1	3.3 ± 0.4
I.C. 95%	]132.9; 137.7[	]66.2; 70.3[	]66.7; 70.3[	]5.3; 5.8[	]284.4; 339.1[	]3.2; 3.4[
Female	121.5 ± 6.1	51.0 ± 2.5	52.2 ± 2.9	6.6 ± 0.4	235.7 ± 70.3	3.9 ± 0.4
I.C. 95%	]118.6; 124.5[	]49.7; 52.4[	]50.6; 53.8[	]6.3; 6.8[	]201.9; 269.7[	]3.6; 4.1[
Cohen's d	−1.82	−3.05	−3.19	1.19	−0.88	1.4
I.C. 95%	]−2.4; −1.2[	]−3.8; −2.2[	]−4.0; −2.3[	]0.6; 1.7[	]−1.4; −0.3[	]0.8; 1.9[

Age-related differences were evident in several of the assessed performance measures, with older players generally outperforming their younger peers (Table 2). Statistically significant differences between age groups were observed in in-water vertical jump height (F(1,62) = 3.725, $p = 0.058$, $\eta^2p = 0.057$), shot velocity without displacement (F(1,52) = 5.864, $p = 0.019$, $\eta^2p = 0.101$), aerobic fitness (F(1,61) = 7.811, $p = 0.007$, $\eta^2p = 0.114$), and change of direction performance (F(1,62) = 7.408, $p = 0.008$, $\eta^2p = 0.107$). These results suggest that age contributes meaningfully to the development of strength, endurance and agility-related capacities in young water polo players. Conversely, no significant age-related effects were found for shot velocity with previous displacement (F(1,53) = 2.124, $p = 0.151$, $\eta^2p = 0.039$) or for 10 m sprint performance (F(1,62) = 0.935, $p = 0.337$, $\eta^2p = 0.015$), indicating that these specific skills may plateau earlier or be less sensitive to age-related physiological development. Furthermore, no significant interaction effects between sex and age were detected in any of the tests, suggesting that the influence of each variable on performance is independent and additive rather than synergistic. Taken together, these findings support the role of age as a contributing factor to certain motor abilities, particularly those involving explosiveness, aerobic capacity and agility, but that this is not uniform across all skill domains.

Table 2. Descriptive statistics (mean ± SD) of specific motor abilities performance by age group and the interaction between sex and age.

Variables	In-Water Vertical Jump (cm)	Shot Velocity with Previous Displacement (km·h^{-1})	Shot Velocity Without Displacement (km·h^{-1})	10 m Sprint (s)	Aerobic Fitness (m)	Change of Direction (s)
Young	129.9 ± 8.5	63.6 ± 8.7	63.3 ± 7.6	5.9 ± 1.2	256.8 ± 82.9	3.5 ± 0.5
I.C. 95%	]126.9; 132.9[	]60.2; 67.0[	]60.3; 66.3[	]5.5; 6.3[	]226.9; 286.8[	]226.9; 286.8[
Adult	132.8 ± 10.9	63.3 ± 10.5	64.3 ± 10.3	5.8 ± 0.7	321.2 ± 91.6	3.4 ± 0.4
I.C. 95%	]128.9; 136.7[	]59.2; 67.3[	]60.4; 68.2[	]5.5; 6.0[	]288.7; 353.7[	]3.2; 3.5[
Interaction Sex–age	$F_{(1,62)} = 1.961$ $p = 0.166$ $\eta^2_p = 0.031$	$F_{(1,53)} = 0.898$ $p = 0.348$ $\eta^2_p = 0.017$	$F_{(1,52)} = 1.399$ $p = 0.242$ $\eta^2_p = 0.026$	$F_{(1,62)} = 0.183$ $p = 0.671$ $\eta^2_p = 0.003$	$F_{(1,62)} = 3.383$ $p = 0.071$ $\eta^2_p = 0.053$	$F_{(1,62)} = 0.761$ $p = 0.386$ $\eta^2_p = 0.012$

4. Discussion

Given the pivotal role of technical skills in water polo performance and the evident inter-team variability, enhancing this competence is essential for improving players' game outcomes [17,24]. To assess water polo players' technical skills, sex- and age-related disparities were examined using specific performance tests. The current study demonstrates that both sex and age significantly influence sport-specific technical skills among water polo players. Key performance indicators, such as in-water vertical jump, swimming velocity, and change of direction ability, showed significant differences between male and female players, as well as across development age groups. These findings align with prior research highlighting physiological differences, including muscle power and anaerobic capacity [25,26], but further contribute by providing sport-specific insights. Notably, younger players exhibited lower velocities and reduced explosiveness, suggesting that standardised training protocols may inadequately address the specific development needs of these subgroups [27].

This underrepresentation in performance research underscores the necessity of individualised skill development approaches. Our data showed established sex- and age-related patterns, but also unexpected deviations. As hypothesised, male players were superior in all specific technical skills tested and showed better aerobic and anaerobic capacity. Older players outperformed their younger counterparts in in-water vertical jump height, non-displacement shot velocity, aerobic capacity and agility, suggesting age-related development in these specific technical skills. However, age was not associated with shot velocity following displacement or 10 m sprint performance, suggesting these skills may plateau earlier or exhibit lower age sensitivity. Sex and age independently influenced performance, as their interaction was not significant.

Water polo offensive and defensive specific technical skills are crucial to the success of a team, as tactical performance depends on their efficient use during the game [28,29]. For instance, vertical jumping ability using the eggbeater kick is crucial for performance [30], enabling high passes, shots over defenders and body position control [19,31]. Our players exhibited lower in-water vertical jump heights compared to male [32] and female [33] elite players. Sex-based disparities were observed, as previous assessments of mean eggbeater force, maximal force, mean peak force, jump distance, and total impulse also revealed disparities between the sexes [34]. Furthermore, in-water vertical jump ability and the velocity at which the ball is thrown are widely recognised as crucial factors in the execution of an effective shot at goal [20,35].

The results observed regarding sex disparities in shooting velocity showed that male players scores were consistent with findings from previous studies [36,37], while the performance of our female players was below average when compared to their counterparts in other studies [25,38,39]. These sex disparities are largely due to differences in muscle mass, upper body strength and power in male players [9]. This also supports prior findings indicating that sex influences playing status in professional water polo [40]. However, female players compensate for physical disparities through refined technique, strategic shot placement, and tactical awareness, highlighting the multidimensional nature of shooting efficiency in elite water polo [16]. These results emphasise the need to include the development of specific technical skills and game intelligence in training, especially for female players, despite physiological limitations.

The physiological demands in water polo require endurance, strength and strategy to sustain intense efforts and minimise fatigue [26]. Aerobic metabolism plays a dominant role, but anaerobic energy is crucial for sprints and physical contact [16,41]. Our female players showed an anaerobic capacity equivalent that of players in the Australian National league, while male players achieved test results comparable to those of the national squad [42]. In addition, the sprint performance of both sexes corresponded to that observed in Spanish water polo players [43]. Previous research shows that players maintain a high heart rate during play, highlighting the need for aerobic fitness for recovery and optimal performance [44,45]. Regarding aerobic capacity, the players of the Australian National Women's Team outperformed our players, in both sexes, illustrating the differences in competition levels [46].

Water polo demands rapid movement transitions and spatial awareness [26]. Our female players outperformed both the Canadian team and comparable male counterparts in change of direction performance [47,48]. Younger males also outperformed older ones [49], highlighting sex and age effects on agility. The current study showed that age influenced the performance specific technical skills. After puberty, maturation, experience and training outweigh physical growth [5], often causing greater variation among players than between age groups [7,50]. In a previous performance factors study with youth water polo players, age-related disparities in specific motor abilities were analysed. Similarly to our data, it was found that older players performed better in motor tasks, though not in swimming velocity [51]. This can be explained by the improvements in coordination during early adolescence, and then individual disparities matter more than growth [52]. These findings suggest age aids physical development, but younger players may need targeted training to improve technique, agility and power.

5. Conclusions

The present study has some limitations that should be acknowledged. First, the unbalanced sex distribution, with a larger number of male players compared to female players. This imbalance may affect the interpretation of sex-related comparisons and limit the generalisability of the findings to female players. Secondly, anthropometric characteristics were not systematically collected or reported for the players, which limits the ability to contextualise performance outcomes and compare them with previous studies. Future research should aim to include more balanced samples to allow stronger conclusions regarding sex differences. However, the findings of this study provide important information for coaches and practitioners who want to optimise water polo training by addressing sex- and age-related disparities in specific technical skills.

Male players were consistently superior to female players in in-water vertical jump, shooting velocity, sprint velocity, aerobic fitness, and change of direction, indicating the need for sex-specific training strategies. Female players could benefit from targeted strength and shooting drills that emphasise technique and tactical awareness to compensate for physiological deficits. Also, coaches and researchers should acknowledge that players' performance is related to age, which implies skill-based training regardless of age. To ensure balanced development, age-specific programs should prioritise foundational technical skills in younger players while progressively incorporating more complex tactical elements. Furthermore, continuous monitoring and adaptation of training loads are essential to address both maturation and individual progression effectively.

Author Contributions: Conceptualisation, F.N., S.C. and R.F.; methodology, F.N., S.C. and R.F.; validation, F.N., S.C. and R.F.; formal analysis, F.N., J.A. and F.C.; investigation, F.N., D.D.C. and A.S.M.; resources, F.N., J.A. and R.F.; data curation, F.N., D.D.C. and A.S.M.; writing—original draft preparation, F.N. and R.F.; writing—review and editing, F.N., F.C., J.A. and R.F.; visualisation, F.N.; supervision, F.N., F.C., J.A. and R.F.; project administration, F.N., S.C. and R.F. All authors have read and agreed to the published version of the manuscript.

Funding: This research received no external funding.

Institutional Review Board Statement: The study was conducted in accordance with the Declaration of Helsinki and approved by the Ethics Committee of Faculty of Sport of University of Porto (CEFADE and 13.2017, 19 January 2017).

Informed Consent Statement: Informed consent was obtained from all subjects involved in the study.

Data Availability Statement: The original contributions presented in the study are included in the article; further inquiries can be directed to the corresponding authors.

Acknowledgments: To the water polo national teams of the Portuguese Swimming Federation.

Conflicts of Interest: The authors declare no conflicts of interest. The funders had no role in the design of the study; in the collection, analyses or interpretation of data; in the writing of the manuscript; or in the decision to publish the results.

References

1. James, J.J.; Hunter, S.K.; Bueckers, E.P.; Joyner, M.J.; Senefeld, J.W. Sex-Based Differences in the Representation of Top Youth Athletes. *Med. Sci. Sports Exerc.* **2025**, *57*, 1523–1529. [CrossRef]
2. Hunter, S.K.; Angadi, S.S.; Bhargava, A.; Harper, J.; Hirschberg, A.L.; Levine, B.D.; Moreau, K.L.; Nokoff, N.J.; Stachenfeld, N.S.; Bermon, S. The Biological Basis of Sex Differences in Athletic Performance: Consensus Statement for the American College of Sports Medicine. *Med. Sci. Sports Exerc.* **2023**, *55*, 2328–2360. [CrossRef] [PubMed]
3. Senefeld, J.W.; Hunter, S.K. Hormonal Basis of Biological Sex Differences in Human Athletic Performance. *Endocrinology* **2024**, *165*, bqae036. [CrossRef] [PubMed]
4. Baxter-Jones, A.D.G.; Barbour-Tuck, E.N.; Dale, D.; Sherar, L.B.; Knight, C.J.; Cumming, S.P.; Ferguson, L.J.; Kowalski, K.C.; Humbert, M.L. The role of growth and maturation during adolescence on team-selection and short-term sports participation. *Ann. Hum. Biol.* **2020**, *47*, 316–323. [CrossRef]
5. Malina, R.M.; Rogol, A.D.; Cumming, S.P.; Coelho E Silva, M.J.; Figueiredo, A.J. Biological maturation of youth athletes: Assessment and implications. *Br. J. Sports Med.* **2015**, *49*, 852–859. [CrossRef]
6. Allen, S.V.; Hopkins, W.G. Age of Peak Competitive Performance of Elite Athletes: A Systematic Review. *Sports Med.* **2015**, *45*, 1431–1441. [CrossRef]
7. Tingelstad, L.M.; Raastad, T.; Till, K.; Luteberget, L.S. The development of physical characteristics in adolescent team sport athletes: A systematic review. *PLoS ONE* **2023**, *18*, e0296181. [CrossRef]
8. Kawai, E.; Tsunokawa, T.; Sakaue, H.; Takagi, H. Propulsive forces on water polo players' feet from eggbeater kicking estimated by pressure distribution analysis. *Sports Biomech.* **2020**, *22*, 1319–1333. [CrossRef]
9. Tsekouras, Y.E.; Kavouras, S.A.; Campagna, A.; Kotsis, Y.P.; Syntosi, S.S.; Papazoglou, K.; Sidossis, L.S. The anthropometrical and physiological characteristics of elite water polo players. *Eur. J. Appl. Physiol.* **2005**, *95*, 35–41. [CrossRef]
10. Garrett, W.E.; Kirkendall, D.T. *Exercise and Sport Science*; Lippincott Williams & Wilkins: Ambler, PA, USA, 2000.
11. Kovačević, N.; Mihanović, F.; Lušić Kalcina, L.; Matijaš, T.; Rukavina, I.; Galić, T. Gender Differences in Cognitive Functions of Youth Water Polo Players. *Sport Mont* **2024**, *22*, 79–85. [CrossRef]
12. Garcia-Marin, P.; Iturriaga, F.M.A. Water polo shot indicators according to the phase of the championship: Medallist versus non-medallist players. *Int. J. Perform. Anal. Sport* **2017**, *17*, 642–655. [CrossRef]
13. Vila, H.; Abraldes, J.A.; Alcaraz, P.E.; Rodriguez, N.; Ferragut, C. Tactical and shooting variables that determine win or loss in top-Level in water polo. *Int. J. Perform. Anal. Sport* **2011**, *11*, 486–498. [CrossRef]
14. Joyner, M.J.; Hunter, S.K.; Senefeld, J.W. Evidence on sex differences in sports performance. *J. Appl. Physiol.* **2025**, *138*, 274–281. [CrossRef]
15. Dong, L.; Bourrel, N.; Paradelo, D.; Christie, S.; Berryman, N.; Romeas, T. The Co-Creation of Representative Training Designs in Elite Water Polo Using a Scientist-Practitioner Approach. *J. Sport Psychol. Action* **2025**. [CrossRef]
16. Tan, F.; Polglaze, T.; Dawson, B. Activity profiles and physical demands of elite women's water polo match play. *J. Sports Sci.* **2009**, *27*, 1095–1104. [CrossRef]
17. McCluskey, L.; Lynskey, S.; Leung, C.K.; Woodhouse, D.; Briffa, K.; Hopper, D. Throwing velocity and jump height in female water polo players: Performance predictors. *J. Sci. Med. Sport* **2010**, *13*, 236–240. [CrossRef]
18. Melchiorri, G.; Triossi, T.; Bianchi, D.; Tancredi, V.; Viero, V. Physical Characteristics and Performance Tests in Male Water Polo: A Multiple Regression Analysis on Youth. *Int. J. Environ. Res. Public Health* **2022**, *19*, 8241. [CrossRef] [PubMed]
19. Platanou, T. On-water and dryland vertical jump in water polo players. *J. Sports Med. Phys. Fit.* **2005**, *45*, 26–31.
20. Canossa, S.; Abraldes, J.A.; Soares, S.M.; Fernandes, R.J.; Garganta, J.M. Vertical jump and shot speed, efficacy and accuracy in water polo. *Int. J. Perform. Anal. Sport* **2016**, *16*, 64–79. [CrossRef]
21. Rechichi, C.; Dawson, B.; Lawrence, S.R. A multistage shuttle swim test to assess aerobic fitness in competitive water polo players. *J. Sci. Med. Sport* **2000**, *3*, 55–64. [CrossRef] [PubMed]
22. Tucher, G.; Castro, F.A.D.; da Silva, A.J.R.M.; Garrido, N.D. Sensitivity and Validity of a Functional Test for Agility Performance in Water Polo Players. *Kinesiology* **2016**, *48*, 124–131. [CrossRef]
23. Cohen, J. *Statistical Power Analysis for the Behavioral Sciences*; L. Erlbaum Associates: Mahwah, NJ, USA, 1988.
24. Hraste, M.; Bulić, K.; Jelaska, I. Differences in Technical and Tactical Efficiency between Winning and Defeated Water Polo Teams: A Comparative Analysis. *Sport Mont* **2023**, *21*, 17–21. [CrossRef]

25. Platanou, T.; Varamenti, E. Relationships between anthropometric and physiological characteristics with throwing velocity and on water jump of female water polo players. *J. Sports Med. Phys. Fit.* **2011**, *51*, 185–193.
26. Botonis, P.G.; Toubekis, A.G.; Platanou, T.I. Physiological and Tactical on-Court Demands of Water Polo. *J. Strength Cond. Res.* **2019**, *33*, 3188–3199. [CrossRef]
27. Perazzetti, A.; Tessitore, A.; Özkol, M.Z.; Novoselac, N.; Dopsaj, M. Assessment of Youth Water Polo Players' Swimming Sprint Potential: A New Approach to Building an International Model. *J. Funct. Morphol. Kinesiol.* **2025**, *10*, 200. [CrossRef] [PubMed]
28. Koopmann, T.; Faber, I.; Baker, J.; Schorer, J. Assessing Technical Skills in Talented Youth Athletes: A Systematic Review. *Sports Med.* **2020**, *50*, 1593–1611. [CrossRef]
29. Platanou, T.; Varamenti, E. Impact of ball possession time and number of passes on the efficiency of scoring in men's water polo. *Hum. Mov.* **2023**, *24*, 18–24. [CrossRef]
30. Nakashima, M.; Minami, Y.; Takagi, H. Optimizing simulation for lower limb motion during throwing in water polo. *Mech. Eng. J.* **2015**, *2*, 14-00472. [CrossRef]
31. Sanders, R.H. Analysis of the eggbeater kick used to maintain height in water polo. *J. Appl. Biomech.* **1999**, *15*, 284–291. [CrossRef]
32. Sáez de Villarreal, E.; Suarez-Arrones, L.; Requena, B.; Haff, G.G.; Ramos Veliz, R. Enhancing performance in professional water polo players: Dryland training, in-water training, and combined training. *J. Strength Cond. Res.* **2015**, *29*, 1089–1097. [CrossRef]
33. Veliz, R.R.; Suarez-Arrones, L.; Requena, B.; Haff, G.G.; Feito, J.; de Villarreal, E.S. Effects of in-Competitive Season Power-Oriented and Heavy Resistance Lower-Body Training on Performance of Elite Female Water Polo Players. *J. Strength Cond. Res.* **2015**, *29*, 458–465. [CrossRef]
34. Croteau, F.; Lanoé, E.; Clément, J. Analysis of Pushing Forces During the Water Polo Eggbeater: Reliability and Validity of a Novel Approach. *J. Sports Sci. Med.* **2024**, *23*, 147–155. [CrossRef]
35. Panero, E.; Agostini, V.; Gastaldi, L. Biomechanical Assessment of Throwing Gesture and Performance in Female Water-Polo Players. *Appl. Sci.* **2022**, *12*, 7856. [CrossRef]
36. Canossa, S.; Abraldes, J.A.; Estriga, L.; Fernandes, R.J.; Garganta, J. Water Polo Shooting Performance: Differences Between World Championship Winning, Drawing and Losing Teams. *J. Hum. Kinet.* **2020**, *72*, 203–214. [CrossRef] [PubMed]
37. Melchiorri, G.; Viero, V.; Triossi, T.; De Sanctis, D.; Padua, E.; Salvati, A.; Galvani, C.; Bonifazi, M.; Del Bianco, R.; Tancredi, V. Water polo throwing velocity and kinematics: Differences between competitive levels in male players. *J. Sports Med. Phys. Fit.* **2015**, *55*, 1265–1271.
38. Abraldes, J.A.; Ferragut, C.; Rodriguez, N.; Vila, M.H. Tactical and shooting variables that determine the difference between win or loss in top-level female water polo: Analysis by phases of the game. *Int. J. Perform. Anal. Sport* **2012**, *12*, 373–384. [CrossRef]
39. Martinez, J.G.; Vila, M.H.; Ferragut, C.; Noguera, M.M.; Abraldes, J.A.; Rodriguez, N.; Freeston, J.; Alcaraz, P.E. Position-Specific Anthropometry and Throwing Velocity of Elite Female Water Polo Players. *J. Strength Cond. Res.* **2015**, *29*, 472–477. [CrossRef]
40. Vila, M.H.; Manchado, C.; Abraldes, J.A.; Ferragut, C. Predicting playing status in professional water polo players: Analysis by gender. *J. Sports Med. Phys. Fit.* **2018**, *58*, 1234–1239. [CrossRef] [PubMed]
41. Platanou, T.; Geladas, N. The influence of game duration and playing position on intensity of exercise during match-play in elite water polo players. *J. Sports Sci.* **2006**, *24*, 1173–1181. [CrossRef]
42. Tan, F.H.Y.; Polglaze, T.; Dawson, B.; Cox, G. Anthropometric and Fitness Characteristics of Elite Australian Female Water Polo Players. *J. Strength Cond. Res.* **2009**, *23*, 1530–1536. [CrossRef] [PubMed]
43. Barrenetxea-García, J.; Nuell, S.; Garai, S.; Murua-Ruiz, A.; Mielgo-Ayuso, J.; Calleja-González, J.; Sáez de Villarreal, E. Effect of Foam Roll recovery method on performance in water polo players: A randomized controlled trial. *Phys. Sportsmed.* **2024**, *52*, 262–270. [CrossRef]
44. Botonis, P.G.; Toubekis, A.G.; Platanou, T.I. Physical Performance During Water-Polo Matches: The Effect of the Players' Competitive Level. *J. Hum. Kinet.* **2016**, *54*, 135–142. [CrossRef]
45. Botonis, P.G.; Toubekis, A.G.; Terzis, G.D.; Geladas, N.D.; Platanou, T.I. Performance Decrement and Skill Deterioration During a Water Polo Game are Linked With the Conditioning Level of the Athletes. *J. Strength Cond. Res.* **2016**, *30*, 1033–1041. [CrossRef]
46. Tan, F.H.Y.; Polglaze, T.; Dawson, B. Comparison of Progressive Maximal Swimming Tests in Elite Female Water Polo Players. *Int. J. Sports Physiol. Perform.* **2009**, *4*, 206–217. [CrossRef] [PubMed]
47. de Castro, C.D.; Tucher, G.; Paixao, D.A.; Vasques, D.M.; Garrido, N.D.; Castro, F.A.D. Agility, vertical jump, and shot velocity of Brazilian water polo players: Correlations and top performances analysis. *J. Sports Med. Phys. Fit.* **2022**, *62*, 757–762. [CrossRef] [PubMed]
48. Dong, L.; Paradelo, D.; Delorme, A.; Oliveira, J.; Parillo, B.; Croteau, F.; Romeas, T.; Dube, E.; Bieuzen, F.; Billaut, F.; et al. Sport-Specific Agility and Change of Direction in Water Polo: The Reliability and Validity of Two Newly Developed Tests. *J. Strength Cond. Res.* **2021**, *35*, S111–S118. [CrossRef]
49. Vasques, D.M.; Tucher, G.; De Castro, C.D.; Paixão, D.A.; Bandeira, P.F.; Castro, F.A.d.S. Anthropometry, Tactical Position, Performance Parameters and Experience in Water Polo: Network Analysis. *Apunt. Educ. Física I Esports* **2023**, *152*, 62–69. [CrossRef]

50. Morganti, G.; Kelly, A.L.; Lascu, A.; Brustio, P.R.; Padua, E.; Filetti, C.; Porta, M.; Briotti, G.; Ruscello, B. Relative age effects in European soccer: Their association with contextual factors, impact on youth national teams' performance, and presence at the senior level. *Front. Sports Act. Living* **2025**, *7*, 1546978. [CrossRef]
51. Noronha, F.; Canossa, S.; Vilas-Boas, J.P.; Afonso, J.; Castro, F.; Fernandes, R.J. Youth Water Polo Performance Determinants: The INEX Study. *Int. J. Environ. Res. Public Health* **2022**, *19*, 4938. [CrossRef]
52. De Siati, F.; Laffaye, G.; Gatta, G.; Dello Iacono, A.; Ardigò, L.P.; Padulo, J. Neuromuscular and technical abilities related to age in water-polo players. *J. Sports Sci.* **2016**, *34*, 1466–1472. [CrossRef]

applied sciences

MDPI

Article

Knee Loading Asymmetries During Descent and Ascent Phases of Squatting After ACL Reconstruction

Manuel Angel Romero Padron [1], Alyx Jorgensen [2], David M. Werner [3], Matthew Alan Tao [1,2] and Elizabeth Wellsandt [1,2,*]

1 Department of Orthopaedic Surgery and Rehabilitation, University of Nebraska Medical Center, 4014 Leavenworth St., Omaha, NE 68105, USA; maromeropadron@volunteer.unmc.edu (M.A.R.P.); matt.tao@unmc.edu (M.A.T.)
2 Physical Therapy Program, University of Nebraska Medical Center, 4014 Leavenworth St., Omaha, NE 68105, USA; alyx.jorgensen@aegistherapies.com
3 College of Allied Health Professions, University of Nebraska Medical Center, 3908 Jones St., Omaha, NE 68198, USA; dwerner@unmc.edu
* Correspondence: elizabeth.wellsandt@unmc.edu

Featured Application

This study shows that squat phase and speed affect interlimb loading asymmetries after ACL reconstruction. Clinicians should monitor movement tempo and target neuromuscular control during rehabilitation to reduce asymmetries and improve outcomes during closed kinetic chain exercises.

Abstract

Asymmetries are common during squats following anterior cruciate ligament reconstruction (ACLR). This study examined interlimb loading differences between squat phases at 6 months post-ACLR. Thirty-five participants performed bodyweight squats at self-selected speed and were analyzed using 3D motion capture. Vertical ground reaction force impulse (vGRFi), external knee flexion moment impulse (KFMi) and hip-to-knee flexion moment impulse ratio (HKRi) were calculated, along with interlimb ratios (ILR). Squat phase durations were also recorded. Paired t-tests and ANCOVA (controlling for time) were used to compare biomechanical variables across squat phases. Greater asymmetry was observed during ascent for vGRFi ILR ($p = 0.045$), KFMi ILR ($p < 0.001$) and HKRi ILR ($p = 0.006$). The ascent phase was faster than descent ($p = 0.036$). After adjusting for time, phase-related differences in ILRs were no longer significant. These findings suggest that greater limb and knee-specific loading asymmetries occur during the ascent phase of squats but may be influenced by movement speed. Importantly, significant knee-specific loading asymmetries persisted regardless of squat phase. At 6 months post-ACLR, addressing neuromuscular control and movement speed during rehabilitation may help reduce biomechanical imbalances during closed kinetic chain exercises.

Keywords: rehabilitation; biomechanics; movement asymmetry; ACL; neuromuscular control

1. Introduction

Squats are a commonly performed exercise during rehabilitation after anterior cruciate ligament (ACL) reconstruction (ACLR) [1,2]. During early- and mid-rehabilitation stages after ACLR, intralimb compensations during squats that shift loading toward the hip and ankle joints of the injured limb, as well as shifted loading to the non-injured limb, occur [3,4]. These compensatory movement patterns often persist after patients have completed rehabilitation, as seen up to 7 years after ACLR [5]. Restoring symmetric movement patterns is important because altered knee joint loading is associated with negative outcomes such as a second ACL injury and the development of knee osteoarthritis (OA) [4,6–8].

Previous research has focused on loading asymmetries during the entire length of squats (descent combined with ascent phase) in individuals after ACLR. A study performed by Sigward et al. compared the injured vs. uninjured limb of individuals after ACLR during the entire length of squats, finding that individuals exhibited lower peak knee flexion angles, knee extension moments and peak vertical ground reaction forces while squatting at 3 and 5 months after surgery [1]. Through investigations of squat biomechanics by Sanford et al. and Garrison et al., the injured knee of individuals after ACLR was compared to healthy controls during the entire length of squats, with increases in anterior femoral translation relative to the tibia, asymmetric ground reaction forces, increased hip energy absorption contributions and decreased knee energy absorption contributions present [5,9]. While previous studies have demonstrated squat-related asymmetries, the distinct phases of the squat [descent and ascent] impose different demands on the musculoskeletal and neuromuscular systems (e.g., knee extensors elongate during descent and shorten during ascent), warranting phase-specific investigation [10,11]. Only the work performed by Strong et al. described squat asymmetries during the descent and ascent phases of squats [4]. However, their data compared the injured knee after ACLR to a single limb of healthy controls without measures of interlimb asymmetries, reporting greater differences in knee flexion and ankle flexion moments during both the descent and ascent phases at 2.9 months and the descent phase alone at 8.8 months post-ACLR ($p < 0.05$) [4]. Nonetheless, the comparison of interlimb squat biomechanics during the descent compared to the ascent phase of squats remains absent from the current literature.

The purpose of this study was to determine if loading symmetries differ between the descent and ascent phases of bilateral squats at 6 months after ACLR and if the length of time during each phase influences squat loading patterns. We hypothesized that interlimb loading asymmetries would be present and similar during the descent and ascent phases of the bodyweight squats. Findings from this study may inform more phase-specific assessment and rehabilitation strategies aimed at reducing persistent movement asymmetries and optimizing knee joint loading after ACLR. Ultimately, reducing asymmetries during functional movements may help decrease the risk of early osteoarthritis and reinjury following primary ACL reconstruction.

2. Materials and Methods

2.1. Study Design

This is a cross-sectional secondary analysis from a prospective cohort study aiming to determine the association between joint loading and cartilage degeneration after ACLR [12]. Participants were enrolled pre-operatively early after sustaining an ACL injury and followed through 18 months after ACLR. The timepoint of interest in this study was 6 months after ACLR.

2.2. Participants

Participants between the ages of 15 and 35 who were within one month of ACL injury at the time of enrollment were eligible for the larger prospective study [12]. Exclusion criteria included previous knee injury or surgery in either knee, history of inflammatory disease, immune compromise, chronic use of NSAIDS, history of cortisone injection during prior three months, current pregnancy, contraindications to MRI, contraindications to

MRI, symptomatic concomitant grade III tear to other knee ligaments requiring surgical intervention, a pre-operatively planned meniscectomy by the treating orthopaedic surgeon and acute or degenerative chondral lesions present on post-injury MRI. All participants provided written informed consent (minors provided assent with written consent from a parent or legal guardian).

Participant Characteristics

Participants self-reported age, sex and race. All study data were collected and/or managed using Research Electronic Data Capture (REDCap) electronic data capture tools hosted at UNMC. REDCap is a secure, web-based application designed to support data capture for research studies [13,14]. REDCap at UNMC is supported by the Research IT Office, funded by the Vice Chancellor for Research (VCR). Operative reports were collected from each participant's surgeon and were used to identify graft type and meniscus repair.

2.3. Squats

Three-dimensional motion analysis was used to examine participants' bilateral squats. Kinematic and kinetic data were simultaneously collected with two embedded force platforms (1080 Hz; Bertec Corporation, Columbus, OH, USA) and an eight-camera motion capture system (120 Hz; Qualisys Inc, Goteborg, Sweden). Passive retroreflective markers (14 mm) were placed on anatomical landmarks of the trunk, pelvis and lower extremities as previously described, including the sternum, C7, T10 and bilateral acromia, upper back, posterior superior iliac spine (PSIS), iliac crest, anterior superior iliac spine (ASIS), medial and lateral femoral epicondyles, medial and lateral malleoli, superior and inferior heel and the first and fifth metatarsal heads [15]. In addition, rigid shells, each containing four retroreflective markers, were secured bilaterally at the lateral thigh and shank.

Participants first completed a 1-s static trial in anatomical position [15]. After this calibration trial, the markers on the ASIS, femoral epicondyles, malleoli and first metatarsal heads were removed prior to dynamic trials. As part of a larger testing protocol (gait, single-leg squats, drop vertical jumps), participants then completed three sets of five bilateral squats with arms crossed over the chest [16]. They were instructed to squat to a comfortable depth at a self-selected speed without further instruction or practice.

Marker trajectories were labeled in Qualisys Track Manager and exported to Visual3D (HAS-Motion, Kingston, ON, Canada). Target and ground reaction force (GRF) data were low-pass filtered using a fourth-order bidirectional Butterworth filter with a 12 Hz cutoff frequency [1].

A subject-specific model including the trunk and lower extremities was created using measured height and mass to estimate segment lengths and joint centers as previously described [1]. To account for the marker diameter and its base, virtual landmarks were offset by 9 mm toward the underlying bone [17]. Virtual ASIS and PSIS landmarks were used to create a Visual 3D composed pelvis. Hip joint centers were estimated using the method of Bell et al., knee joint centers were defined as the midpoints between the virtual femoral epicondyle landmarks, and ankle joint centers were defined as the midpoints between the virtual malleolar landmarks [18,19].

Each squat was segmented into descent and ascent phases. The start of the squat was defined when vertical center-of-mass velocity exceeded 0.02 m/s downward, and the end of the squat was defined when upward center-of-mass velocity decreased below 0.02 m/s (Figure 1). The lowest vertical center-of-mass position determined the transition point between the descent and ascent phases. External knee and hip moments were calculated using an inverse dynamics approach, and joint impulses for the descent and ascent phases were computed using the trapezoidal rule [20].

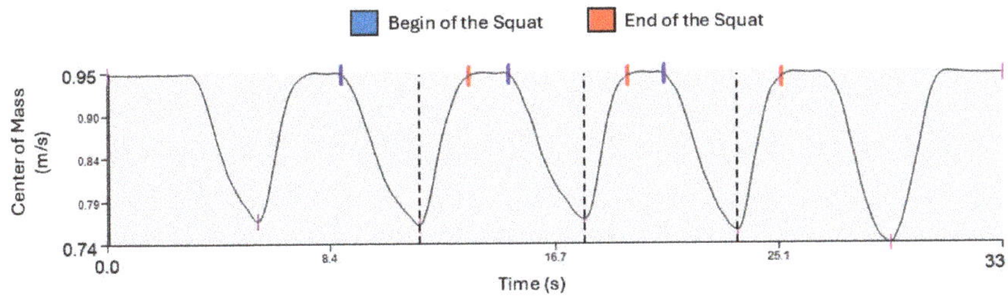

Figure 1. Graphical representation of the body's center of mass (COM) in the vertical direction. The vertical velocity of the COM was used to define the beginning and end of each squat. The middle three of each set of five squats were used for analysis. Dashed lines represent the end of the descent phase and the beginning of the ascent phase, defined by the lowest vertical position of the COM. Abbreviations: m/s, meter per second; s. seconds.

Variables of interest included vertical ground reaction force impulse (vGRFi), external knee flexion moment impulse (KFMi), ratio of external hip flexion to knee flexion moment impulse (hip-to-knee ratio (HKRi)) and time spent in the descent and ascent phase. Data from the 3 middle squats (second, third and fourth) of each set of squats (9 total squats averaged) was averaged to account for more variable movement patterns during the first and fifth squats of each set [1]. For each variable, an interlimb ratio (ILR) was calculated using this equation: involved limb/uninvolved limb. An ILR of 1 indicates an equal distribution of load between limbs (vGRFi), between knees (KFMi) or between the hip and knee joints across limbs (HKRi). An ILR of less than 1 for vGRFi and KFMi indicates lower loading in the injured limb or knee compared to the uninjured limb or knee. An ILR greater than 1 for the HKRi indicates a greater contribution of loading in the hip than the knee in the injured limb compared to the uninjured limb. Time spent during the descent and ascent phases was expressed in seconds.

2.4. Statistical Analysis

Statistical analyses were completed using IBM SPSS Statistics (IBM Corp. Released 2022. IBM SPSS Statistics for Windows, Version 30.0. IBM Corp: Armonk, NY, USA). Demographic and surgical characteristics were described using means and standard deviations or frequencies and percentages, as appropriate. First, paired t-tests were performed to compare each biomechanical variable (vGRFi, KFMi, HKRi) between phases. One outlier for HKRi was detected that was more than 1.5 box-lengths from the edge of the box in a boxplot. It represented expected variation within the sample population and thus was kept in the analysis. Three extreme outliers for HKRi were detected that were more than 3.0 box-lengths from the edge of the box in a boxplot. These 3 outliers were removed from the analysis of HKRi due to their leverage on statistical results. Assumptions of normality were not violated, as assessed by visual inspection of histograms and P-P plots. Secondly, an ANCOVA was used to determine if ILRs for each squat variable differed between phases after controlling for time. There was a linear relationship between time on each phase and each of the biomechanical variables for each of the phases, as assessed by visual inspection of a scatterplot. Standardized residuals were normally distributed, as assessed by Shapiro–Wilk's test ($p > 0.05$). There was homoscedasticity and homogeneity of variances, as assessed by visual inspection of a scatterplot and Levene's test of homogeneity of variance, respectively. There were no outliers in the data, as assessed by no cases with standardized residuals greater than ± 3 standard deviations. A p-value of 0.05 was set a priori.

3. Results

Forty participants were initially enrolled in the study. Five participants were lost to follow-up by 6 months after ACLR, resulting in thirty-five participants included in this analysis. Descriptive statistics for sex, age, race, weight, height and BMI of the included participants are presented in Table 1. More than half of the participants (54%) had a concomitant meniscal repair. Nearly two-thirds (63%) of participants had a patellar tendon autograft during ACLR. Overall limb loading (vGRFi) was more symmetric compared to knee-specific loading contributions (KFMi, HKRi) (Table 2, Figure 2).

Table 1. Participant and surgical characteristics.

	Number of Participants (%) or Mean (SD)	95% Confidence Interval
Sex (female)	21 (60.0)	
Age (years)	19.7 (4.6)	18.1–21.3
Race		
Asian	2 (5.7)	
Black or African American	3 (8.6)	
Hispanic, Latino or Spanish	3 (8.6)	
White	25 (71.4)	
Other	2 (5.7)	
Weight (kg)	74.2 (16.3)	68.6–79.8
Height (m)	1.7 (0.1)	1.7–1.7
Body mass index (BMI; kg/m^2)	25.9 (4.8)	24.2–27.5
Concomitant meniscus repair (Yes)	19 (54.3)	
Graft type		
Hamstring tendon autograft	4 (11.4)	
Patellar tendon autograft	22 (62.9)	
Quadriceps tendon autograft	8 (22.9)	
Allograft	1 (2.9)	

Abbreviations: %, percentage; SD, standard deviation; kg, kilograms; m, meters; BMI, body mass index.

Table 2. Results of interlimb ratios for vertical ground reaction force impulse, external knee flexion moment impulse and ratio of external hip flexion to knee flexion moment impulse, as well as time during the different phases of squats.

	Descent (Mean ± SD)			Ascent (Mean ± SD)			*p*-Value
	Injured	Uninjured	ILR	Injured	Uninjured	ILR	
vGRFi (BW/kg·m)	0.48 ± 0.12	0.52 ± 0.13	0.93 ± 0.12	0.45 ± 0.10	0.51 ± 0.10	0.90 ± 0.14	0.045 *
KFMi (N·m·s/kg·m)	0.22 ± 0.10	0.35 ± 0.15	0.70 ± 0.39	0.18 ± 0.09	0.32 ± 0.11	0.64 ± 0.39	<0.001 *
HKRi (N·m·s/kg·m)	1.74 ± 0.71	1.12 ± 0.53	1.84 ± 0.93	2.04 ± 0.84	1.17 ± 0.53	2.06 ± 1.12	0.006
Time (s)		1.00 ± 0.25			0.96 ± 019		0.036 *

* *p*-value of <0.05. Abbreviations: SD, standard deviation; ILR, interlimb ratio; vGRFi, vertical ground reaction force impulse; BW, bodyweight; s, seconds; kg, kilograms; m, meters; KFMi, external knee flexion moment impulse; N, newtons; HKRi, ratio of external hip flexion to knee flexion moment impulse.

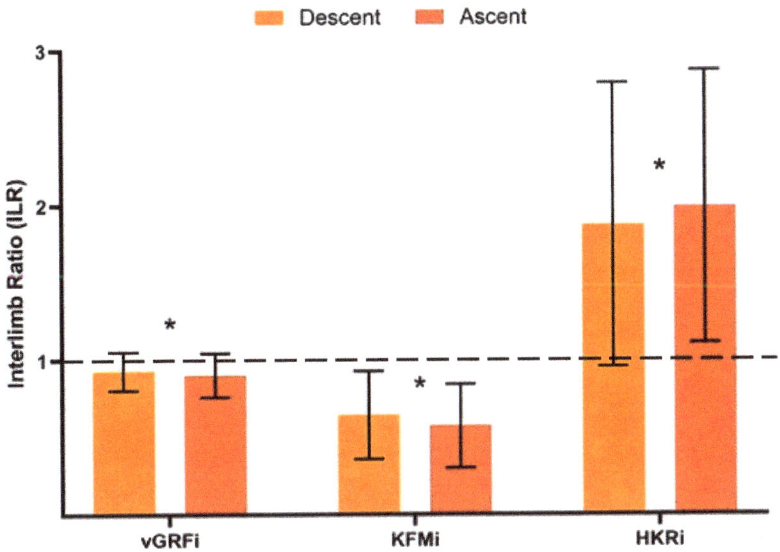

Figure 2. Interlimb ratios for vertical ground reaction force impulse, external knee flexion moment impulse and ratio of external hip flexion to knee flexion moment impulse during the descent and ascent phases of squats. Error bars represent ±1 standard deviation. * p-value of <0.05. Note: The dashed line represents an interlimb ratio of 1 and an equal distribution of loading for each variable between the injured and uninjured limb. Abbreviations: ILR, interlimb ratio; vGRFi, vertical ground reaction force impulse; KFMi, external knee flexion moment impulse; HKRi, ratio of external hip flexion to knee flexion moment impulse.

There was a significant difference in loading asymmetry between the descent and ascent phases of bilateral squats for all biomechanical variables (Table 3). Participants demonstrated more asymmetry during the ascent phase for vGRFi ILR ($p = 0.045$; Descent (D): 0.93 ± 0.12, Ascent (A): 0.90 ± 0.14), KFMi ILR ($p < 0.005$; D: 0.70 ± 0.39, A: 0.64 ± 0.39) and HKRi ILR ($p = 0.006$; D: 1.84 ± 0.93, A: 2.06 ± 1.12) (Figure 1). There was a significant difference in time with the ascent phase being faster than the descent phase ($p = 0.036$; D:1.00 ± 0.25 s, A:0.96 ± 0.19 s) (Figure 2).

Table 3. Results of paired t-tests comparing the vertical ground reaction force impulse, external knee flexion moment impulse, ratio of external hip flexion to knee flexion moment impulse and time in each squat phase during the descent compared to ascent phase of bilateral squats.

		Mean Difference	SD	SEM	95% CI	p
vGRFi ILR	Unadjusted	0.029	0.083	0.014	0.001 to 0.058	0.045 *
	Adjusted	0.024	-	0.032	−0.039 to 0.088	0.449
KFMi ILR	Unadjusted	0.066	0.095	0.016	0.034 to 0.099	<0.005 *
	Adjusted	0.071	-	0.093	−0.115 to 0.257	0.449
HKRi ILR †	Unadjusted	−0.218	0.418	0.074	−0.368 to −0.067	0.006 *
	Adjusted	−0.223	-	0.260	−0.744 to 0.298	0.395
Time (s)	-	0.047	0.129	0.022	0.003 to 0.092	0.036 *

* p-value of <0.05. † Three extreme outliers were not included in this analysis ($n = 32$). Abbreviations: SD, standard deviation; SEM, standard error of the mean; s, seconds; t, t-value; df, degrees of freedom; vGRFi, vertical ground reaction force impulse: KFMi, external knee flexion moment impulse; HKRi, ratio of external hip flexion to knee flexion moment impuse; ILR, interlimb ratio.

After adjusting for time during each phase, there was no longer a statistically significant difference in vGRFi ILR ($F(1, 67) = 0.581$, $p = 0.449$, partial $\eta^2 = 0.009$), KFMi ILR ($F(1, 67) = 0.579$, $p = 0.449$, partial $\eta^2 = 0.009$) or HKRi ILR ($F(1, 61) = 0.733$, $p = 0.395$, partial $\eta^2 = 0.012$). Adjusted means and variability for each of the biomechanical variables are provided in Table 3.

4. Discussion

Our results indicate that at 6 months after ACLR, participants continue to exhibit deficits in knee loading symmetry during both the descent and ascent phases of squatting. Initially, greater asymmetry was observed during the ascent phase, as evidenced by vGRFi, KFMi and HKRi, compared to the descent phase. Specifically, the injured limb demonstrated approximately 30% lower external knee flexion moment impulse compared to the uninjured limb, along with nearly twice the hip flexion moment contribution relative to knee flexion moment contribution (HKRi). However, given that the ascent phase was performed at a faster speed than the descent (4% faster), we controlled for movement time to account for potential confounding. After adjusting for time, interlimb loading asymmetries persisted, but the previously observed difference in asymmetry between the ascent and descent phases was no longer statistically significant.

The findings of our study align with and expand upon previous research. Sigward et al., Sanford et al. and Garrison et al. reported asymmetries when analyzing the squat as a single movement, while Strong et al. identified phase-specific asymmetries when comparing the limb of participants after ACLR to a healthy control limb [1,4,5,9]. In our study, asymmetries were evident in both phases; however, they were more pronounced during the ascent phase, which may be attributed to several factors. First, participants were instructed to perform squat sets without standardized control over time under tension and phase-specific velocity. Our findings indicate that differences in movement velocity across phases may influence an individual's control over the squatting movement, particularly during the ascent phase, thereby highlighting pre-existing asymmetrical movement patterns. Additionally, neuromuscular control of the lower limb and muscle activation are critical factors in squat execution during rehabilitation after ACLR. It is well established that neural control patterns differ across eccentric compared to concentric muscle contractions, with a lower rate of motor unit recruitment required under equal force conditions during eccentric versus concentric muscle contraction [21]. ACL injury and subsequent ACLR often result in arthrogenic muscle inhibition (AMI), a neural impairment that restricts proper quadriceps activation in the affected limb that is still present up to 2 years after ACLR [22,23]. Given that greater quadriceps activation is required during the ascent phase compared to the descent phase of a bodyweight squat, reduced quadriceps neuromuscular activation may contribute to the greater asymmetry observed in this phase [24]. Finally, the depth of the squats was selected at the discretion of the participants according to their comfort and ability. Deeper squats have been associated with increased muscle activation, particularly in the quadriceps and gluteus maximus [25]. Therefore, variations in squat depth, along with potential improper muscle activation, may contribute to greater asymmetries during the ascent phase compared to the descent phase.

Strengths and Limitations

This is the first study to compare loading asymmetries between the injured and uninjured limbs during the descent and ascent phases of bilateral squats after ACLR. By breaking down the movement into its directional components, we gain novel insight into the specific biomechanical strategies patients use after ACLR, which could inform more targeted rehabilitation protocols to address persistent deficits. Limitations of our study include a relatively small sample size of athletic participants with a young average age (19 years). Therefore, our findings may not be generalized to older populations. In addition, the differences in vGRFi, KFMi and HKRi interlimb ratios between the descent phase and ascent phase were relatively small; it is unknown if these differences are clinically meaningful. Finally, not controlling squat depth across participants aided in the generalizability of our study findings to real-world movements but could have impacted findings.

5. Conclusions

At 6 months after ACLR, interlimb asymmetries persist during both the descent and ascent phases of the squat, with greater asymmetries observed during the ascent phase. However, these asymmetries were no longer present after controlling for phase duration. This suggests that underlying neuromuscular control deficits remain even during controlled bodyweight activities at this stage of the rehabilitation process. Furthermore, movement speed during rehabilitation exercises should be carefully considered to effectively mitigate biomechanical imbalances in closed kinetic chain movements.

Author Contributions: Conceptualization and Study Design: all authors; Data Collection and Investigation: A.J., D.M.W. and E.W.; Data Analysis and Interpretation: all authors.; Writing—Original Draft and Preparation: M.A.R.P. and E.W.; Writing—Review and Editing: all authors. All authors have read and agreed to the published version of the manuscript.

Funding: This work was supported in part by the Foundation for Physical Therapy Research (New Investigator Fellowship Training Initiative), the Rheumatology Research Foundation (Investigator Award), the National Institutes of Health (R21AR075254; 2U54GM115458) and internal funding from the University of Nebraska Medical Center.

Institutional Review Board Statement: The study was conducted according to the guidelines of the Declaration of Helsinki and approved by the Institutional Review Board of the University of Nebraska Medical Center (UNMC) (IRB#215-20-EP; 07-19-2024—approved on 15 June 2023).

Informed Consent Statement: Informed consent was obtained from all subjects involved in the study.

Data Availability Statement: The data that support the findings of this study are available from the corresponding author upon reasonable request. The data are not publicly available due to privacy.

Acknowledgments: We would like to thank Robert Barber for his assistance with study coordination and data collection and Ryan Arnold and Kellen Huston for their assistance with participant recruitment.

Conflicts of Interest: The authors have no professional relationships with companies or manufacturers that will benefit from the results of the present study. Other conflicts include the following: E.W. is funded by U54GM115458, R01AR080346, R15AG085105, R34AR083077 and the Arthritis Foundation Osteoarthritis Clinical Trials Network (OACTN) Clinical Trial Unit and serves with roles for the following: Osteoarthritis Research Society International (Rehabilitation Discuss Group Steering Committee) and East Omaha Athletic Association (Board Member). M.T. serves on the editorial board for Current Review in Musculoskeletal Medicine, provides consulting for NewClip and Vericel and owns stock in Overture and NeAT Surgical. For the remaining authors, no conflicts were declared.

References

1. Sigward, S.M.; Chan, M.S.; Lin, P.E.; Almansouri, S.Y.; Pratt, K.A. Compensatory strategies that reduce knee extensor demand during a bilateral squat change from 3 to 5 months following anterior cruciate ligament reconstruction. *J. Orthop. Sports Phys. Ther.* **2018**, *48*, 713–718. [CrossRef] [PubMed]
2. Malempati, C.; Jurjans, J.; Noehren, B.; Ireland, M.L.; Johnson, D.L. Current rehabilitation concepts for anterior cruciate ligament surgery in athletes-pubmed. *Orthopedics* **2015**, *38*, 689–696. [CrossRef] [PubMed]

3. Garrison, J.C.; Hannon, J.; Goto, S.; Kosmopoulos, V.; Aryal, S.; Bush, C.; Bothwell, J.M.; Singleton, S.B. Knee loading after acl-r is related to quadriceps strength and knee extension differences across the continuum of care. *Orthop. J. Sports Med.* **2019**, *7*, 2325967119870155. [CrossRef]
4. Strong, A.; Markström, J.L.; Schelin, L.; Häger, C.K. Asymmetric loading strategies during squats following anterior cruciate ligament reconstruction: A longitudinal investigation with curve analyses throughout and after rehabilitation-pubmed. *Scand. J. Med. Sci. Sports* **2024**, *34*, e14524. [CrossRef] [PubMed]
5. Sanford, B.A.; Williams, J.L.; Zucker-Levin, A.; Mihalko, W.M. Asymmetric ground reaction forces and knee kinematics during squat after anterior cruciate ligament (acl) reconstruction. *Knee* **2016**, *23*, 820–825. [CrossRef]
6. Wellsandt, E.; Gardinier, E.S.; Manal, K.; Axe, M.J.; Buchanan, T.S.; Snyder-Mackler, L. Decreased knee joint loading associated with early knee osteoarthritis after anterior cruciate ligament injury. *Am. J. Sports Med.* **2015**, *44*, 143–151. [CrossRef]
7. Paterno, M.V.; Schmitt, L.C.; Ford, K.R.; Rauh, M.J.; Myer, G.D.; Huang, B.; Hewett, T.E. Biomechanical measures during landing and postural stability predict second anterior cruciate ligament injury after anterior cruciate ligament reconstruction and return to sport. *Am. J. Sports Med.* **2010**, *38*, 1968–1978. [CrossRef]
8. Renström, P.; Arms, S.W.; Stanwyck, T.S.; Johnson, R.J.; Pope, M. Strain within the anterior cruciate ligament during hamstring and quadriceps activity. *Am. J. Sports Med.* **1986**, *14*, 83–87. [CrossRef]
9. Garrison, J.C.; Hannon, J.; Goto, S.; Giesler, L.; Bush, C.; Bothwell, J.M. Participants at three months post-operative anterior cruciate ligament reconstruction (acl-r) demonstrate differences in lower extremity energy absorption contribution and quadriceps strength compared to healthy controls. *Knee* **2018**, *25*, 782–789. [CrossRef]
10. Myer, G.D.; Kushner, A.M.; Brent, J.L.; Schoenfeld, B.J.; Hugentobler, J.; Lloyd, R.S.; Vermeil, A.; Chu, D.A.; Harbin, J.; McGill, S.M. The back squat: A proposed assessment of functional deficits and technical factors that limit performance. *Strength Cond. J.* **2014**, *36*, 4–27. [CrossRef]
11. Straub, R.K.; Powers, C.M. A biomechanical review of the squat exercise: Implications for clinical practice-pubmed. *Int. J. Sports Phys. Ther.* **2024**, *19*, 490. [CrossRef]
12. Wellsandt, E.; Manzer, M.; Sajja, B.; Golightly, Y.M.; Wichman, C.; Tao, M. Cumulative knee joint loading and early markers of cartilage microstructure after acl injury. *Osteoarthr. Cartil.* **2024**, *32*, S278–S279. [CrossRef]
13. Harris, P.A.; Taylor, R.; Minor, B.L.; Elliott, V.; Fernandez, M.; O'Neal, L.; McLeod, L.; Delacqua, G.; Delacqua, F.; Kirby, J.; et al. The redcap consortium: Building an international community of software platform partners. *J. Biomed. Inform.* **2019**, *95*, 103208. [CrossRef]
14. Harris, P.A.; Taylor, R.; Thielke, R.; Payne, J.; Gonzalez, N.; Conde, J.G. Research electronic data capture (redcap)—A metadata-driven methodology and workflow process for providing translational research informatics support. *J. Biomed. Inform.* **2009**, *42*, 377–381. [CrossRef] [PubMed]
15. Wellsandt, E.; Kallman, T.; Golightly, Y.; Podsiadlo, D.; Dudley, A.; Vas, S.; Michaud, K.; Tao, M.; Sajja, B.; Manzer, M. Knee joint unloading and daily physical activity associate with cartilage T_2 relaxation times 1 month after acl injury. *J. Orthop. Res.* **2021**, *40*, 138–149. [CrossRef] [PubMed]
16. Kritz, M.; Cronin, J.; Hume, P. The bodyweight squat: A movement screen for the squat pattern. *Strength Cond. J.* **2009**, *31*, 76–85. [CrossRef]
17. Fiorentino, N.M.; Kutschke, M.J.; Atkins, P.R.; Foreman, K.B.; Kapron, A.L.; Anderson, A.E. Accuracy of functional and predictive methods to calculate the hip joint center in young non-pathologic asymptomatic adults with dual fluoroscopy as a reference standard. *Ann. Biomed. Eng.* **2016**, *44*, 2168–2180. [CrossRef]
18. Bell, A.L.; Brand, R.A.; Pedersen, D.R. Prediction of hip joint centre location from external landmarks. *Hum. Mov. Sci.* **1989**, *8*, 3–16. [CrossRef]
19. Bell, A.L.; Pedersen, D.R.; Brand, R.A. A comparison of the accuracy of several hip center location prediction methods. *J. Biomech.* **1990**, *23*, 617–621. [CrossRef]
20. Winter, D.A. *Biomechanics and Motor Control of Human Movement*, 4th ed.; John Wiley & Sons: Hoboken, NJ, USA, 2009; pp. 107–138.
21. Duchateau, J.; Baudry, S. Insights into the neural control of eccentric contractions. *J. Appl. Physiol.* **2014**, *116*, 1418–1425. [CrossRef]
22. Patel, H.H.; Berlinberg, E.J.; Nwachukwu, B.; Williams, R.J.; Mandelbaum, B.; Sonkin, K.; Forsythe, B. Quadriceps weakness is associated with neuroplastic changes within specific corticospinal pathways and brain areas after anterior cruciate ligament reconstruction: Theoretical utility of motor imagery-based brain-computer interface technology for rehabilitation. *Arthrosc. Sports Med. Rehabil.* **2023**, *5*, e207–e216. [PubMed]
23. Hunnicutt, J.L.; McLeod, M.M.; Slone, H.S.; Gregory, C.M. Quadriceps neuromuscular and physical function after anterior cruciate ligament reconstruction. *J. Athl. Train.* **2020**, *55*, 238–245. [CrossRef] [PubMed]

24. Wilk, K.E.; Escamilla, R.F.; Fleisig, G.S.; Barrentine, S.W.; Andrews, J.R.; Boyd, M.L. A comparison of tibiofemoral joint forces and electromyographic activit during open and closed kinetic chain exercises. *Am. J. Sports Med.* **1996**, *24*, 518–527. [CrossRef] [PubMed]
25. O'nEill, K.E.; Psycharakis, S.G. The effect of back squat depth and load on lower body muscle activity in group exercise participants. *Sports Biomech.* **2021**, *23*, 555–566. [CrossRef]

applied sciences

MDPI

Article

Profiling of Physical Qualities of Highly Trained Portuguese Youth Soccer Players

Miguel Silva [1,2], Hugo Duarte Antunes [2,3], Ana Sousa [1], Fábio Yuzo Nakamura [1], António Rodrigues Sampaio [1,4] and Ricardo Pimenta [2,3,4,5,*]

1 Research Center in Sports Sciences, Health Sciences and Human Development (CIDESD), University of Maia (UMaia), 4475-690 Maia, Portugal; mlpsmiguel@gmail.com (M.S.); acsousa@umaia.pt (A.S.); fnakamura@umaia.pt (F.Y.N.); arsampaio@umaia.pt (A.R.S.)
2 Department of Rehabilitation and Optimization of Performance (DROP), Futebol Clube Famalicão-Futebol SAD, 4760-482 Famalicão, Portugal; hugoduarteantunes@gmail.com
3 CIPER, Faculdade de Motricidade Humana, Universidade de Lisboa, 1495-751 Cruz Quebrada Dafundo, Portugal
4 Research Center of the Polytechnic Institute of Maia (N2i), Maia Polytechnic Institute (IPMAIA), Castêlo da Maia, 4475-690 Maia, Portugal
5 Porto Biomechanics Laboratory, Faculty of Sport, University of Porto, 4200-450 Porto, Portugal
* Correspondence: rjl.pimenta@gmail.com

Abstract: Background: A comprehensive understanding of variations in physical attributes both between and within young players is crucial for accurately identifying talent based on physical performance. This study aimed to compare maximum strength, jump, and sprint test results among young soccer players from different age categories and playing positions. Secondarily, this study aimed to analyze the association between maximum strength, jump, and sprint performances. Methods: A total of 103 players were categorized as U23, U19, U17, and U16. The players were placed into these age categories based on their football abilities. All participants completed standardized warm-ups, and testing procedures followed protocols established in previous studies. Results: Significant differences were found between age categories regarding the Isometric Mid-Thigh Pull ($p < 0.001$; $\eta^2_p = 0.33$), Countermovement Jump (CMJ) ($p < 0.001$; $\eta^2_p = 0.50$), Squat Jump (SJ) ($p < 0.001$; $\eta^2_p = 0.29$), and sprint ($p < 0.001$; $\eta^2_p = 0.30$) tests. No significant differences were detected in Broad Jump results between age categories. Moreover, no significant differences were observed in any physical capacities between playing positions. Furthermore, significant moderate-to-strong correlations ($r = 0.30$–0.86) were observed between all physical tests. Conclusions: Age categories can distinguish soccer players' performance in different physical tests while no differences were observed between playing positions. Similar magnitude correlations were observed across all tests with only the CMJ and SJ being strongly correlated. Coaches and strength and conditioning professionals should apply a variety of tests to assess different physical qualities since they have different patterns between age categories.

Keywords: strength; athlete profiling; age category; playing position; maximal voluntary isometric contraction; soccer

1. Introduction

The profiling of soccer players highlights physical, physiological, and anthropometric characteristics [1–3]. Previous studies have reported significant differences regarding running performance and strength capacities between players of distinct age categories and playing positions [4,5]. Indeed, elite soccer players have shown better physical performance compared to their non-elite counterparts, whether at senior level or in youth categories [6]. This information is highly valuable for coaches and sports scientists since it enables the assessment and comparison of players' physical capacities, regardless of their playing position or competition level [1]. Consequently, it supports the process of individualization during training prescriptions and load monitoring without disregarding the natural predisposition of some players to exhibit greater or lesser capacity in specific performance metrics [7]. In soccer, the profiling of players is intimately related to the profile of games and training practices, especially concerning the player's field position and age category. Indeed, significant differences between playing positions and age categories in soccer players across different physical qualities were previously reported [8,9], demonstrating the necessity of coaches to individualize training prescriptions.

Soccer is characterized as a high-intensity interval-based sport, which requires a blend of skills such as speed, agility, endurance, and strength [10]. The most decisive actions in soccer depend on maximal strength and anaerobic power generated by the neuromuscular system, which is directly linked to the capacity for force production [11]. It has been suggested that the Isometric Mid-Thigh Pull (IMTP) test is effective for assessing maximum strength due to its validity, reliability, simplicity, and low risk [12]. Previous studies have reported that isometric tests such as the IMTP are well-suited for daily soccer routines since it results in minimal muscle damage compared to eccentric-based tests [13], which are commonly used for stratifying strength training programs, and serves as an index of fatigue [14]. Moreover, IMTP performance has been correlated with sprint performance, which is highly correlated with peak speed in professional English youth (U23) soccer players [12]. IMTP absolute peak force has shown greater differences in higher age categories, with moderate to large effect sizes (d = 0.62 to 1.66), and the largest effects observed in the highest age groups [15].

Jumping and sprinting abilities are widely used as indicators of neuromuscular fitness in youth players and as talent identification markers to differentiate between elite and non-elite soccer players [16]. Although jumping movements are not the most frequent in soccer matches, analytical testing involves lower technical complexity compared to sprinting and change of direction movements while providing distinct performance parameters. Vertical jumping, in particular, reflects lower-limb muscular explosive strength and is useful for monitoring training-induced adaptations [17,18]. Common exercises for evaluating rapid strength include the Countermovement Jump (CMJ) and Squat Jump (SJ) [19]. It has been reported that stronger players (measured via the IMTP) have demonstrated better CMJ performance [12].

Sprinting is another critical action that can determine the outcome of a soccer game. Indeed, straight sprints are the most dominant actions in scoring situations, with the majority performed without the ball, emphasizing the importance of assessing sprint performance [20,21]. Previous studies reported significant differences between age categories in sprinting performance [22]. Moreover, it has been shown that elite young soccer players are faster, stronger, and show better jumping performance than their non-elite counterparts [23]. Additionally, among elite young soccer players, noticeable differences between field positions are reported regarding various performance metrics including sprint, favoring higher performances of offensive position players (e.g., wide midfielders and forwards) [24,25].

Therefore, a comprehension of physical qualities between young players is essential for effectively recognizing talent based on their physical performance attributes and optimizing training strategies according to each player's profile. This study aims to (i) compare the physical capacities of young soccer players from a professional club across different age categories; and (ii) compare the physical capacities between playing positions. Secondly the present study aims to analyze whether strength, jump, and sprint tests are correlated. The hypotheses are as follows: (i) higher age categories will perform better across all tests; (ii) forwards will achieve superior sprint test results compared to their peers in other positions; and (iii) IMTP and vertical jump tests will be correlated.

2. Methods

One hundred and three highly trained male soccer players from a Portuguese club, were recruited to participate in this study, from U23 (n = 24 players; age: 19.7 ± 1.6 yrs; height: 1.80 ± 0.07 m; body mass: 75.6 ± 6.9 kg), U19 (n = 28 players; age: 18.1 ± 0.6 yrs; height: 1.79 ± 0.06 m; body mass: 72.2 ± 6.8 kg), U17 (n = 25 players; age: 16.6 ± 0.6 yrs; height: 1.76 ± 0.06 m; body mass: 67.9 ± 6.2 kg), and U16 (n = 26 players; age: 15.5 ± 0.5 yrs; height: 1.72 ± 0.06 m; body mass: 65.2 ± 8.6 kg). Regarding exclusion criteria, only goalkeepers and injured players, who were not able to be assessed in these tests, were excluded, in order to ensure a more representative number of subjects. Data collection was part of the teams' routines in which players are assessed throughout the season. Therefore, normal ethics committee clearance was not required [26]. However, the club's director provided written consent for the data for the study.

2.1. Dynamometry

The Piforce had two load cells (CATLT DYLY103) of 200 kg, each with a maximum capacity of 1961 N (one on each side to measure the force applied by each leg). The force was measured at a sampling rate of 80 Hz using an Arduino Uno with an HX711 converter. For the IMTP (Figure 1A), participants were placed in a standing position and the bar height was adjusted up or down to allow the athlete to obtain the optimal knee (125–145°) and hip (140–150°) angles [27]. Participants were instructed to exert force "as fast and strong as possible" to obtain peak force (PF) [28]. The PF was estimated with Python code using Visual Studio (Microsoft Corporation, Redmond, WA, USA). The onset of force production was defined by visual detection as previously suggested [28] for each maximal voluntary contraction (MVC) and quantified in Newtons. After that, all MVCs were processed using an automated routine in Visual Studio Code, which gave a Peak Force value for both cells. For data analysis, maximum repetition was selected for each test, and the average was calculated between the two cells for those tests.

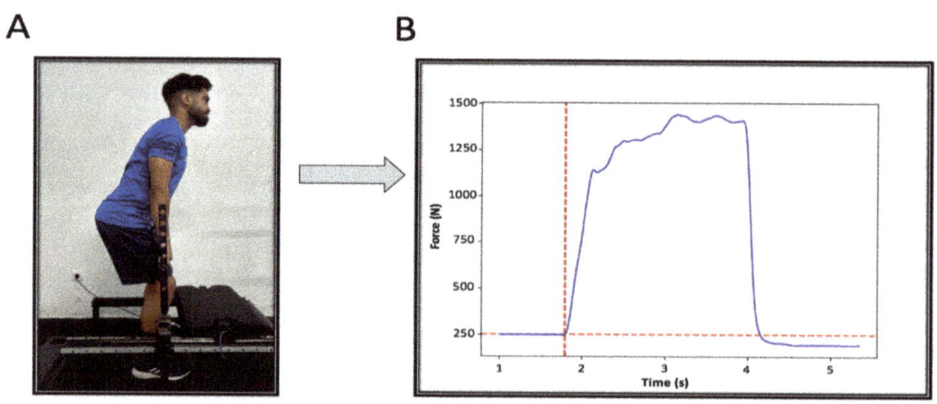

Figure 1. (**A**) Isometric Mid-Thigh Pull (IMTP). (**B**) Force–time curve of the IMTP exercise using a Python routine implemented in Visual Studio Code, the blue line represents the force-time curve, while the red line indicates the onset of force production.

2.2. Jumps and Speed Equipment

The CMJ, SJ, and DJ tests were assessed by Chronojump software (Chronojump 2.3.0-83), which has been used to assess jump metrics [29]. The Broad Jump (BJ) was assessed using a tape measure with participants being positioned with the front part of the foot on the 0 m mark. The sprint test was assessed by using the same Chronojump software (Chronojump 2.3.0-83) with two photocells that were placed at the 0 m mark and at the end mark (40 m), with the participant being positioned 1 m behind the photocell. These procedures were previously used to assess the speed of football players [1,30].

2.3. Protocols

Soccer players were familiarized with each exercise in previous training sessions. The tests were conducted under the same conditions with a 48 h recovery post-game. To increase the tests' reliability, players performed the same warm-up protocol, which consisted of a 5 min exercise on a stationary bike. To minimize the fatigue effect between tests, the assessments were split into three sessions: one for the maximum strength test (IMTP), another for rapid strength tests (CMJ, SJ, DJ, and BJ), and a third for the sprint test. The order of the tests in each session was randomized. Regarding the IMTP, after the standard warm-up, players performed two maximal voluntary isometric contractions (MVICs) of 3 s with a 2 min recovery between trials. If a countermovement occurred prior to the beginning of the pull (IMTP), the trial was not considered, and the player was asked to repeat it. Moreover, in relation to the CMJ, SJ, DJ, and BJ, after the standard warm-up, players performed two maximal jumps with a 1 min recovery between reps. The DJ was performed from a 20 cm high box. For all jumps, if hip flexion, dorsiflexion, or an arm swing occurred along the jump, the trial was not considered, and the player was asked to repeat it. For the sprint test, after a general warm-up, players performed a single 40 m sprint after resting for 2 min. For all tests (with the exception of the IMTP), the best repetition was used in the analysis.

2.4. Statistical Analysis

All data were analyzed using IBM SPSS Statistics 27.0 (IBM Corporation, Armonk, NY, USA). An assumption test was first conducted to check for data normality, outliers, and homogeneity of variance-covariance matrices, with no violations identified. To analyze differences between age categories and playing positions, a one-way ANOVA was performed. Post-hoc analysis was conducted using Bonferroni correction to determine differences within each factor. Moreover, the correlation between tests were analyzed using Pearson's and Spearman's coefficient for normally and not normally distributed data, respectively. The correlation coefficients were classified as weak (<0.3), moderate (0.3–0.7), and strong (>0.7) [31]. The partial eta squared values were reported as a measure of the effect size of the ANOVA findings, and were classified as small (η^2_p = 0.01–0.05), medium (η^2_p = 0.06–0.013), and large ($\eta^2_p > 0.14$) effects. Additionally, the significance was set at $p < 0.05$, and to analyze the effect size, it was also calculated by Cohen's d.

3. Results

3.1. Playing Position

Comparing test results by playing position, no statistically significant differences were found: IMTP ($p = 0.41$; $d = 0.04$), CMJ ($p = 0.26$; $d = 0.06$), SJ ($p = 0.11$; $d = 0.08$), DJ ($p = 0.35$; $d = 0.05$), BJ ($p = 0.33$; $d = 0.05$), and sprint ($p = 0.47$; $d = 0.04$).

3.2. Age Categories

The results revealed that there were statistically significant differences in the IMTP ($p < 0.001$; $\eta^2_p = 0.33$), CMJ ($p < 0.001$; $\eta^2_p = 0.495$), SJ ($p < 0.001$; $\eta^2_p = 0.29$), and sprint ($p < 0.001$; $\eta^2_p = 0.23$) test results among all age categories.

Regarding IMTP performance (Figure 2A), significant differences were found in the following comparisons: U23 and U17 (U23 = 1706 ± 269 N; U17 = 1386 ± 199 N; $p = 0.001$; $d = 1.68$); U23 and U16 (U23 = 1706 ± 269 N; U16 = 1330 ± 228 N; $p < 0.001$; $d = 1.51$); and U19 and U16 (U19 = 1565 ± 176 N; U16 = 1330 ± 228 N; p = 0.007; $d = 1.15$).

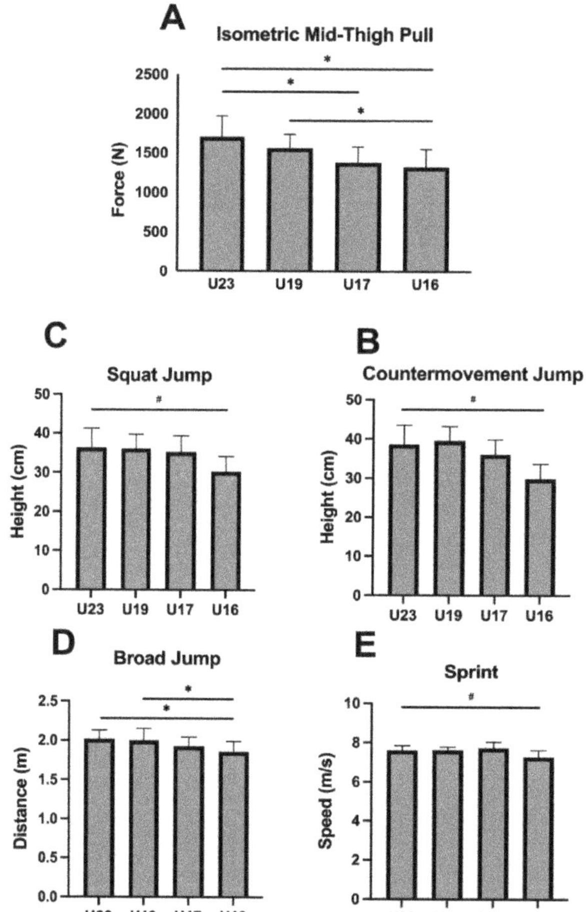

Figure 2. Comparison of test performances between age groups. (**A**) Isometric Mid-Thigh Pull performance in Newtons with significant differences (*) between U23 and U16 ($p = 0.001$), U23 and U17 ($p < 0.001$), and U19 and 16 ($p = 0.007$); (**B**) Countermovement Jump height with significant differences (#) between U23 and U16 ($p < 0.001$), U19 and U16 ($p < 0.001$), and U17 and U16 ($p = 0.003$); (**C**) Squat Jump height with significant (#) differences between U23 and U16 ($p < 0.001$), U19 and U16 ($p < 0.001$), and U17 and U16 ($p < 0.001$). (**D**) Broad Jump distance with significant differences (*) between U23 and U16 ($p = 0.002$), and U19 and U16 ($p = 0.008$); (**E**) sprint speed with significant differences (#) between U23 and U16 ($p = 0.002$), U19 and U16 ($p < 0.001$), and U17 and U16 ($p < 0.001$).

For the SJ (Figure 2B), significant differences were detected between U23 and U16 (U23 = 36.26 ± 4.98 cm; U16 = 30.26 ± 3.87 cm; $p < 0.001$; $d = 2.17$); U19 and U16 (U19 = 36.11 ± 3.57 cm; U16 = 30.26 ± 3.87 cm; $p < 0.001$; $d = 1.57$); and U17 and U16 (U17 = 35.21 ± 4.14; U16 = 30.26 ± 3.87 cm; $p = 0.003$; $d = 1.24$).

In relation to the CMJ (Figure 2C), there were also significant differences in the following comparisons: U23-U16 (U23 = 38.60 ± 4.88 cm; U16 = 29.84 ± 3.83 cm; $p < 0.001$; $d = 2.00$); U19-U16 (U19 = 39.53 ± 3.68 cm; U16 = 29.84 ± 3.83 cm; $p < 0.001$; $d = 3.56$); and U17-U16 (U17 = 36.03 ± 3.81 cm; U16 = 29.84 ± 3.83 cm; $p < 0.001$; $d = 1.62$)

For the BJ (Figure 2D), the only differences that were found were when comparing the following: U23-U16 (U23 = 2.02 ± 0.11 cm; U16 = 1.86 ± 0.13 cm; $p = 0.002$; $d = 2.01$); and U19-U16 (U19 = 2.00 ± 0.15 cm; U16 = 1.86 ± 0.13 cm; $p = 0.008$; $d = 12.76$).

Regarding the sprint (Figure 2E), there were significant differences in the following comparisons: U23-U16 (U23 = 7.61 ± 0.24 m/s; U16 = 7.28 ± 0.35 m/s; $p = 0.002$; $d = 1.10$); U19-U16 (U19 = 7.62 ± 0.16 m/s; U16 = 7.28 ± 0.35 m/s; $p < 0.001$; $d = 1.30$); and U17-U16 (U17 = 7.73 ± 0.31; U16 = 7.28 ± 0.35 m/s; $p < 0.001$; $d = 1.37$).

3.3. Correlation Between Tests

Correlation analysis revealed that all physical performance tests were significantly correlated. Detailed information regarding the strength of the correlation and level of significance between tests can be found in Table 1.

Table 1. Pearson correlation analysis between the different physical performance tests. * denotes a significant correlation between tests ($p < 0.05$).

	IMTP	Squat Jump	Countermovement Jump	Broad Jump	Sprint
IMTP		$r = 0.30$ ($p = 0.003$) *	$r = 0.43$ ($p < 0.001$) *	$r = 0.38$ ($p < 0.001$) *	$r = 0.33$ ($p = 0.002$) *
Squat Jump			$r = 0.86$ ($p < 0.001$) *	$r = 0.59$ ($p < 0.001$) *	$r = 0.37$ ($p < 0.001$) *
Countermovement Jump				$r = 0.60$ ($p < 0.001$) *	$r = 0.42$ ($p < 0.001$) *
Broad Jump					$r = 0.39$ ($p < 0.001$) *
Sprint					

4. Discussion

To the best of our knowledge, this is the first study in soccer that compares physical assessment test results across different age categories and between playing positions, and correlates the physical tests. The main findings of this study were (i) higher age categories outperformed lower age categories in all tests; (ii) no differences were found between playing positions; and (iii) moderate correlations were seen between the IMTP and all the other tests while a strong correlation was observed between the SJ and CMJ.

In accordance with our initial hypothesis, players from higher age categories achieved generally better results. Indeed, significant differences were found on the IMTP between U23 and U17 and U16, and U19 and U16. These findings could be explained by size-dependent factors, such as muscle mass, as well as size-independent factors like genetics and hormonal influences [32]. Players from higher age categories are also more experienced in strength training, benefiting from the morphological and neural adaptations associated with the recurring strength training stimulus, resulting in a greater maximum force capacity.

This is corroborated by studies showing an increase in peak force, measured via IMTP testing, with age throughout different age categories [15,33].

Regarding explosive force, statistically significant differences were found across all tests. The CMJ and SJ tests revealed differences between every age category compared to U16. Moreover, in the BJ, the U23 and U19 age categories performed better compared to the U16 age category. Jumping test performance provides shorter than necessary timeframes for players to achieve their maximal force capacity, and stronger players perform better in jump performance [34]. Stronger players present greater overall force generation capacity independent of the time constraints associated with force production. In fact, late rate of force development and maximum voluntary contraction torque were found to be correlated with muscle mass having a significant influence on the late rate of force development capacity [35]. However previous studies have reported divergent findings. For instance, Castagna and Castellini [19] found that vertical jump was not a suitable metric for distinguishing young Italian national players across different age categories (U17, U20, and U21). The results of the present study are consistent with this observation as the U23, U19, and U17 groups did not exhibit statistically significant differences in vertical jump performance. This lack of differentiation can be attributed to the average ages of the groups examined: 19.7 years for U23 and 18.1 years for U19, which approximately correspond to the U20 category in Castagna and Castellini's study [19], and 16.6 years for U17, aligning closely with their U17 category. This age similarity could explain the similar performance, particularly between the U23 and U19 age categories.

Regarding sprint performance, significant differences were observed in all age categories outperforming the U16. Faster sprint speeds are associated with greater ground reaction forces [36], which faster players present as superior rapid force capacity. Indeed, players from higher age categories presented superior explosive force performance. This can also be attributed to the distinct maturational status of players since different maturational stages lead to specific physiological and structural changes [37–39]. Similar results were reported in two top-level Brazilian clubs in which senior and U20 players presented higher speeds in a linear 20 m sprint compared to U17 and U15 players [40]; significant age-related differences were observed in sprint performance among young soccer players as well [41].

Contrary to our initial hypothesis, no differences were found between playing positions, with previous studies reporting contrasting findings [1,42–46]. For example, greater sprint performance in forwards compared to defenders was observed across different sprint distances (5 m, 10 m, 20 m, and 30 m) [42,43]. On the other hand, no differences in 10 m and 30 m sprints between playing positions were reported [44–46]. Authors suggested that approaching the professional level at higher age categories, positional differences occur with forwards becoming the fastest players of the team [47]. One possible explanation could be due to modern soccer's increasing emphasis on outfield players contributing to overall play, leading to more uniform physical performance across positions [48]. Moreover, a recent study reported no significant differences in sprint performance, knee flexor force capacity, and mechanical properties of the hamstrings between playing positions in players from different competition levels [1]. This finding suggests that such a phenomenon is not exclusive to youth development but also persists in senior competition contexts.

Regarding correlation between tests, a significantly stronger correlation was verified between the CMJ and SJ ($r = 0.86$; $p < 0.001$) due to similarities between the two tests since players applied vertical force on the ground (i.e., vertical force vector). The IMTP also revealed a moderate correlation with explosive jumping and sprinting. Previous studies demonstrated a moderate-to-strong correlation between the IMTP and sprint ($r = -0.69$; $p \leq 0.01$) and jumping performance ($r = 0.81$; $p \leq 0.01$), respectively [49,50].

Although stronger players tend to perform better in jumping and sprint performance tests, the moderate correlations found suggest that other factors besides peak torque are determinant for jumping and sprint performance such as neuromuscular coordination and rapid force development [51]. Both the CMJ and SJ were moderately correlated ($r = 0.60$; $p < 0.001$) with the BJ. Although these tests encompass a squatting pattern, the SJ and CMJ are essentially vertical-oriented while the BJ has a horizontal force component mediating jump performance [52].

Sprinting and jumping can be categorized as explosive movements. The results of the present study indicated that jumping tests were moderately correlated with sprint performance ($r = 0.37$–0.42; $p < 0.001$). However, sprinting and jumping performance present kinetic and kinematic differences due to reliance on distinct attributes. A previous systematic review and meta-analysis demonstrated moderate-to-strong correlations ($r = 0.45$–0.73; $p < 0.001$) between standing long jump distance and acceleration (0–30 m) and maximal sprint (30–100 m), respectively [53]. Moreover, moderate correlations were reported between the SJ ($r = 0.56$; $p < 0.05$) and CMJ ($r = 0.55$; $p < 0.05$) and sprint [54]. In fact, sprinting is associated with a capacity to generate higher ground force reactions, with a specific orientation, in minimal contact times [55], whereas jumps involving a horizontal force vector component (e.g., BJ) are characterized by higher ground contact times, which influences the magnitude of correlations. Indeed, it is expected that players exhibiting greater maximal force production (i.e., higher peak force) generally demonstrate superior outcomes across various performance metrics [56–58]. This is primarily attributable to their elevated relative strength (maximum force capacity relative to body mass), which enhances their ability to overcome inertia more efficiently. Additionally, the development of maximal strength, as well as jumping and sprinting capabilities, appears to exert a mutually reinforcing effect, contributing positively to overall athletic performance. This interrelationship is further supported by the significant correlations frequently observed among distinct physical qualities.

In the context of optimizing specific performance attributes in elite youth soccer players, it is suggested for coaches to employ training strategies that incorporate a broader range of exercises. Such an approach may elicit complementary adaptations beyond those directly targeted through the performance-specific training task itself. Furthermore, future research should seek to replicate the current methodological framework across diverse cohorts of elite youth soccer players. This is particularly relevant given that the criteria defining "elite" status may vary between competitive leagues, potentially influencing the physical performance profiles observed.

Furthermore, establishing normative performance benchmarks for youth soccer players using physical assessments may offer valuable reference values for developing evidence-based return-to-play criteria for common injuries in soccer, such as hamstring strains and anterior cruciate ligament injuries [59,60]. Future studies should monitor the progression of performance in the aforementioned tests alongside the incidence rate of these injuries in order to assess their validity.

This study has some limitations. Firstly, the data were collected at a single time point, and it is known that physical performance can fluctuate over the course of a season. However, it could be extremely challenging to evaluate players across all age categories within the same period using an extensive battery of tests without compromising the success of high-level competitions. Secondly, the fact that the club provides strength training to all players may have influenced the results, making them less generalizable as strength training can either enhance or diminish performance outcomes. Finally, the players were grouped by age categories (competitive level) rather than maturational age, which is naturally an important factor to consider. However, this decision was based on a more ecologically valid and realistic approach. Indeed, in the current context, an increasing number of young players are joining teams with older age groups, including at the senior level.

Further research is needed to examine the relationship between chronological age and competition level through physical assessments that include goalkeepers for a more complete analysis.

5. Conclusions

The results of this study showed that age categories can distinguish performance in physical tests with the highest differences being observed between the U23 and U16 categories. It is likely that older players display overall higher performance levels given the effects of maturity and strength training on maximum strength, rapid strength, and speed. However, these results should be considered context-specific. On the other hand, playing position was shown to not be influential to distinguish players physically. However, it should be noted that these results could be task-specific since soccer is a multidisciplinary sport that does not only depend on physical capacity. Moreover, moderate correlations were observed among the different tests, with the CMJ and SJ showing strong correlations, possibly due the vertical applied force vector in both tests.

The present study indicates that the physical characteristics of players from different age categories and from different positions are becoming more similar, possibly due to specific training characteristics and the influence of physical attributes in modern soccer. Finally, it suggested the implementation of different tests to assess physical qualities as demonstrated in the present study. However, if possible, these tests should be applied at multiple points throughout the season, allowing for the identification not only of the athletes' physical development but also of potential gaps within each age category. This would enable improved monitoring and control of training load.

Author Contributions: Conceptualization, R.P. and M.S.; methodology, R.P. and M.S.; software, R.P.; validation, R.P., H.D.A., F.Y.N. and A.S.; formal analysis, R.P. and H.D.A.; investigation, R.P. and M.S.; resources, R.P. and A.R.S.; data curation, R.P. and M.S.; writing—original draft preparation, R.P. and M.S.; writing—review and editing, R.P., H.D.A., F.Y.N. and A.S.; visualization, R.P., F.Y.N., A.R.S. and A.S.; supervision, R.P., F.Y.N. and A.S.; project administration, R.P.; funding acquisition, R.P. and A.R.S. All authors have read and agreed to the published version of the manuscript.

Funding: This research received no external funding.

Institutional Review Board Statement: Ethical review and approval were waived for this study, as the data were collected retrospectively as part of the team's routine assessments conducted throughout the season.

Informed Consent Statement: Written informed consent has been obtained from the club to publish this paper.

Data Availability Statement: The data that support the findings of this study are available from the corresponding author upon reasonable request.

Acknowledgments: The authors thank the study participants for their participation.

Conflicts of Interest: The authors declare no conflicts of interest.

References

1. Pimenta, R.; Antunes, H.; Bruno, P.; Veloso, A.P. Hamstrings Mechanical Properties Profiling in Football Players of Different Competitive Levels and Positions after a Repeated Sprint Protocol. *Front. Physiol.* **2023**, *14*, 1315564. [CrossRef] [PubMed]
2. Gissis, I.; Papadopoulos, C.; Kalapotharakos, V.I.; Sotiropoulos, A.; Komsis, G.; Manolopoulos, E. Strength and Speed Characteristics of Elite, Subelite, and Recreational Young Soccer Players. *Res. Sports Med.* **2006**, *14*, 205–214. [CrossRef] [PubMed]

3. França, C.; Ihle, A.; Marques, A.; Sarmento, H.; Martins, F.; Henriques, R.; Gouveia, É.R. Physical Development Differences between Professional Soccer Players from Different Competitive Levels. *Appl. Sci.* **2022**, *12*, 7343. [CrossRef]
4. Rebelo, A.; Brito, J.; Maia, J.; Coelho-e-Silva, M.J.; Figueiredo, A.J.; Bangsbo, J.; Malina, R.M.; Seabra, A. Anthropometric Characteristics, Physical Fitness and Technical Performance of under-19 Soccer Players by Competitive Level and Field Position. *Int. J. Sports Med.* **2013**, *34*, 312–317. [CrossRef]
5. Dragula, L.; Lehnert, M.; Psotta, R.; Gonosová, Z.; Valenta, S.; Štastný, P. The Relative Force in Squat Jump Is the Best Laboratory Predictor of Sprint Performance in Adolescent Soccer Players. *Hum. Mov.* **2017**, *2017*, 83–90. [CrossRef]
6. Tereso, D.; Paulo, R.; Petrica, J.; Duarte-Mendes, P.; Gamonales, J.M.; Ibáñez, S.J. Assessment of Body Composition, Lower Limbs Power, and Anaerobic Power of Senior Soccer Players in Portugal: Differences According to the Competitive Level. *Int. J. Environ. Res. Public Health* **2021**, *18*, 8069. [CrossRef]
7. Petr, M.; Thiel, D.; Kateřina, K.; Brož, P.; Malý, T.; Zahálka, F.; Vostatková, P.; Wilk, M.; Chycki, J.; Stastny, P. Speed and Power-Related Gene Polymorphisms Associated with Playing Position in Elite Soccer Players. *Biol. Sport* **2022**, *39*, 355–366. [CrossRef]
8. Marques, M.C.; Izquierdo, M.; Gabbett, T.J.; Travassos, B.; Branquinho, L.; van den Tillaar, R. Physical Fitness Profile of Competitive Young Soccer Players: Determination of Positional Differences. *Int. J. Sports Sci. Coach.* **2016**, *11*, 693–701. [CrossRef]
9. Nobari, H.; Eken, Ö.; Prieto-González, P.; Brito, J.P.; Oliveira, R. Associations among Maturity, Accumulated Workload, Physiological, and Body Composition Factors in Youth Soccer Players: A Comparison between Playing Positions. *Biology* **2022**, *11*, 1605. [CrossRef]
10. Iaia, F.M.; Rampinini, E.; Bangsbo, J. High-Intensity Training in Football. *Int. J. Sports Physiol. Perform.* **2009**, *4*, 291–306. [CrossRef]
11. Hüseyin, H.; Kemalettin, G. The Relationship between Isokinetic Strength and Anaerobic Performance in Elite Youth Football Players. *J. ROL Sports Sci.* **2023**, *4*, 457–468.
12. Mason, L.; Kirkland, A.; Steele, J.; Wright, J. The Relationship between Isometric Mid-Thigh Pull Variables and Athletic Performance Measures: Empirical Study of English Professional Soccer Players and Meta-Analysis of Extant Literature. *J. Sports Med. Phys. Fit.* **2021**, *61*, 645–655. [CrossRef] [PubMed]
13. Hody, S.; Croisier, J.-L.; Bury, T.; Rogister, B.; Leprince, P. Eccentric Muscle Contractions: Risks and Benefits. *Front. Physiol.* **2019**, *10*, 536. [CrossRef]
14. Norris, D.; Joyce, D.; Siegler, J.; Clock, J.; Lovell, R. Recovery of Force-Time Characteristics After Australian Rules Football Matches: Examining the Utility of the Isometric Midthigh Pull. *Int. J. Sports Physiol. Perform.* **2019**, *14*, 765–770. [CrossRef]
15. Morris, R.O.; Jones, B.; Myers, T.; Lake, J.; Emmonds, S.; Clarke, N.D.; Singleton, D.; Ellis, M.; Till, K. Isometric Midthigh Pull Characteristics in Elite Youth Male Soccer Players: Comparisons by Age and Maturity Offset. *J. Strength Cond. Res.* **2020**, *34*, 2947–2955. [CrossRef]
16. Asadi, A.; Ramirez-Campillo, R.; Arazi, H.; Sáez de Villarreal, E. The Effects of Maturation on Jumping Ability and Sprint Adaptations to Plyometric Training in Youth Soccer Players. *J. Sports Sci.* **2018**, *36*, 2405–2411. [CrossRef]
17. Chelly, M.S.; Ghenem, M.A.; Abid, K.; Hermassi, S.; Tabka, Z.; Shephard, R.J. Effects of in-Season Short-Term Plyometric Training Program on Leg Power, Jump- and Sprint Performance of Soccer Players. *J. Strength Cond. Res.* **2010**, *24*, 2670–2676. [CrossRef]
18. Caldwell, B.P.; Peters, D.M. Seasonal Variation in Physiological Fitness of a Semiprofessional Soccer Team. *J. Strength Cond. Res.* **2009**, *23*, 1370–1377. [CrossRef]
19. Castagna, C.; Castellini, E. Vertical Jump Performance in Italian Male and Female National Team Soccer Players. *J. Strength Cond. Res.* **2013**, *27*, 1156–1161. [CrossRef]
20. Faude, O.; Koch, T.; Meyer, T. Straight Sprinting Is the Most Frequent Action in Goal Situations in Professional Football. *J. Sports Sci.* **2012**, *30*, 625–631. [CrossRef]
21. Martínez-Hernández, D.; Quinn, M.; Jones, P. Linear Advancing Actions Followed by Deceleration and Turn Are the Most Common Movements Preceding Goals in Male Professional Soccer. *Sci. Med. Footb.* **2023**, *7*, 25–33. [CrossRef] [PubMed]
22. Reilly, T.; Williams, A.M.; Nevill, A.; Franks, A. A Multidisciplinary Approach to Talent Identification in Soccer. *J. Sports Sci.* **2000**, *18*, 695–702. [CrossRef] [PubMed]
23. Trecroci, A.; Milanović, Z.; Frontini, M.; Iaia, F.M.; Alberti, G. Physical Performance Comparison between Under 15 Elite and Sub-Elite Soccer Players. *J. Hum. Kinet.* **2018**, *61*, 209–216. [CrossRef]
24. Abbott, W.; Brickley, G.; Smeeton, N.J. Physical Demands of Playing Position within English Premier League Academy Soccer. *J. Hum. Sport Exerc.* **2018**, *13*, 285–295. [CrossRef]
25. Pettersen, S.A.; Brenn, T. Activity Profiles by Position in Youth Elite Soccer Players in Official Matches. *Sports Med. Int. Open* **2019**, *3*, E19–E24.
26. Winter, E.M.; Maughan, R.J. Requirements for Ethics Approvals. *J. Sports Sci.* **2009**, *27*, 985. [CrossRef]
27. Comfort, P.; Dos'Santos, T.; Beckham, G.K.; Stone, M.H.; Guppy, S.N.; Haff, G.G. Standardization and Methodological Considerations for the Isometric Midthigh Pull. *Strength Cond. J.* **2019**, *41*, 57–79. [CrossRef]

28. Maffiuletti, N.A.; Aagaard, P.; Blazevich, A.J.; Folland, J.; Tillin, N.; Duchateau, J. Rate of Force Development: Physiological and Methodological Considerations. *Eur. J. Appl. Physiol.* **2016**, *116*, 1091–1116. [CrossRef]
29. Pueo, B.; Penichet-Tomas, A.; Jimenez-Olmedo, J.M. Reliability and Validity of the Chronojump Open-Source Jump Mat System. *Biol. Sport* **2020**, *37*, 255–259. [CrossRef]
30. Pimenta, R.; Lopes, T.; Bruno, P.; Veloso, A. Effects of Repeated Sprints on Hamstring Active Shear Modulus Pattern and Neuromuscular Parameters in Football Players with and without Hamstring Strain Injury History—A Retrospective Study. *NATO Adv. Sci. Inst. Ser. E Appl. Sci.* **2023**, *13*, 3099. [CrossRef]
31. Sheskin, D.J. *Handbook of Parametric and Nonparametric Statistical Procedures*, 2nd ed.; Chapman and Hall/CRC: Boca Raton, FL, USA, 2000; ISBN 9781584881339.
32. Tumkur Anil Kumar, N.; Oliver, J.L.; Lloyd, R.S.; Pedley, J.S.; Radnor, J.M. The Influence of Growth, Maturation and Resistance Training on Muscle-Tendon and Neuromuscular Adaptations: A Narrative Review. *Sports* **2021**, *9*, 59. [CrossRef] [PubMed]
33. Emmonds, S.; Morris, R.; Murray, E.; Robinson, C.; Turner, L.; Jones, B. The Influence of Age and Maturity Status on the Maximum and Explosive Strength Characteristics of Elite Youth Female Soccer Players. *Sci. Med. Footb.* **2017**, *1*, 209–215. [CrossRef]
34. Comfort, P.; Stewart, A.; Bloom, L.; Clarkson, B. Relationships between Strength, Sprint, and Jump Performance in Well-Trained Youth Soccer Players. *J. Strength Cond. Res.* **2014**, *28*, 173–177. [CrossRef] [PubMed]
35. Andersen, L.L.; Aagaard, P. Influence of Maximal Muscle Strength and Intrinsic Muscle Contractile Properties on Contractile Rate of Force Development. *Eur. J. Appl. Physiol.* **2006**, *96*, 46–52. [CrossRef]
36. Weyand, P.G.; Sternlight, D.B.; Bellizzi, M.J.; Wright, S. Faster Top Running Speeds Are Achieved with Greater Ground Forces Not More Rapid Leg Movements. *J. Appl. Physiol. (1985)* **2000**, *89*, 1991–1999. [CrossRef]
37. Malina, R.M.; Rogol, A.D.; Cumming, S.P.; Coelho e Silva, M.J.; Figueiredo, A.J. Biological Maturation of Youth Athletes: Assessment and Implications. *Br. J. Sports Med.* **2015**, *49*, 852–859. [CrossRef]
38. Otero-Esquina, C.; de Hoyo Lora, M.; Gonzalo-Skok, Ó.; Domínguez-Cobo, S.; Sánchez, H. Is Strength-Training Frequency a Key Factor to Develop Performance Adaptations in Young Elite Soccer Players? *Eur. J. Sport Sci.* **2017**, *17*, 1241–1251. [CrossRef]
39. Lloyd, R.S.; Oliver, J.L.; Myer, G.D.; De Ste Croix, M.; Read, P.J. Seasonal Variation in Neuromuscular Control in Young Male Soccer Players. *Phys. Ther. Sport* **2020**, *42*, 33–39. [CrossRef]
40. Loturco, I.; Jeffreys, I.; Abad, C.C.C.; Kobal, R.; Zanetti, V.; Pereira, L.A.; Nimphius, S. Change-of-Direction, Speed and Jump Performance in Soccer Players: A Comparison across Different Age-Categories. *J. Sports Sci.* **2020**, *38*, 1279–1285. [CrossRef]
41. Milanović, Z.; Sporiš, G.; Weston, D.; Kovačević, M.; Pantelić, G. Age-Related Differences in Linear Sprint and Power Characteristics in Youth Elite Soccer Players. *Eur. J. Phys. Educ. Sport Sci.* **2015**, *4*, 132–141.
42. Gil, S.; Ruiz, F.; Irazusta, A.; Gil, J.; Irazusta, J. Selection of Young Soccer Players in Terms of Anthropometric and Physiological Factors. *J Sports Med. Phys. Fitness* **2007**, *47*, 25–32. [PubMed]
43. Tasmektepligil, M.Y.; Arslan, O.; Ermis, E. The Evaluation of Anaerobic Power Values and Sprint Performances of Football Players Playing in Different Positions. *Anthropologist* **2016**, *23*, 497–504. [CrossRef]
44. Lago-Peñas, C.; Casais, L.; Dellal, A.; Rey, E.; Domínguez, E. Anthropometric and Physiological Characteristics of Young Soccer Players according to Their Playing Positions: Relevance for Competition Success. *J. Strength Cond. Res.* **2011**, *25*, 3358–3367. [CrossRef] [PubMed]
45. Wong, P.-L.; Chamari, K.; Dellal, A.; Wisløff, U. Relationship between Anthropometric and Physiological Characteristics in Youth Soccer Players. *J. Strength Cond. Res.* **2009**, *23*, 1204–1210. [CrossRef]
46. Taskin, H. Evaluating Sprinting Ability, Density of Acceleration, and Speed Dribbling Ability of Professional Soccer Players with Respect to Their Positions. *J. Strength Cond. Res.* **2008**, *22*, 1481–1486. [CrossRef]
47. Slimani, M.; Nikolaidis, P.T. Anthropometric and Physiological Characteristics of Male Soccer Players according to Their Competitive Level, Playing Position and Age Group: A Systematic Review. *J. Sports Med. Phys. Fitness* **2019**, *59*, 141–163. [CrossRef]
48. Lago-Peñas, C.; Lorenzo-Martinez, M.; López-Del Campo, R.; Resta, R.; Rey, E. Evolution of Physical and Technical Parameters in the Spanish 2012–2019. *Sci. Med. Footb.* **2023**, *7*, 41–46. [CrossRef]
49. Thomas, C.; Comfort, P.; Chiang, C.-Y.; Jones, P.A. Relationship between Isometric Mid-Thigh Pull Variables and Sprint and Change of Direction Performance in Collegiate Athletes. *J. Trainol.* **2015**, *4*, 6–10. [CrossRef]
50. Townsend, J.R.; Bender, D.; Vantrease, W.C.; Hudy, J.; Huet, K.; Williamson, C.; Bechke, E.; Serafini, P.R.; Mangine, G.T. Isometric Midthigh Pull Performance Is Associated with Athletic Performance and Sprinting Kinetics in Division I Men and Women's Basketball Players. *J. Strength Cond. Res.* **2019**, *33*, 2665–2673. [CrossRef]
51. Van Hooren, B.; Zolotarjova, J. The Difference Between Countermovement and Squat Jump Performances: A Review of Underlying Mechanisms With Practical Applications. *J. Strength Cond. Res.* **2017**, *31*, 2011–2020. [CrossRef]
52. Cronin, J.B.; Brughelli, M.; Gamble, P.; Brown, S.R.; Mckenzie, C. Acute Kinematic and Kinetic Augmentation in Horizontal Jump Performance Using Haltere Type Handheld Loading. *J. Strength Cond. Res.* **2014**, *28*, 1559–1564. [CrossRef] [PubMed]

53. Lin, J.; Shen, J.; Zhang, J.; Zhou, A.; Guo, W. Correlations between Horizontal Jump and Sprint Acceleration and Maximal Speed Performance: A Systematic Review and Meta-Analysis. *PeerJ* **2023**, *11*, e14650. [CrossRef]
54. Kale, M.; Aşçi, A.; Bayrak, C.; Açikada, C. Relationships among Jumping Performances and Sprint Parameters during Maximum Speed Phase in Sprinters. *J. Strength Cond. Res.* **2009**, *23*, 2272–2279. [CrossRef]
55. Morin, J.-B.; Edouard, P.; Samozino, P. Technical Ability of Force Application as a Determinant Factor of Sprint Performance. *Med. Sci. Sports Exerc.* **2011**, *43*, 1680–1688. [CrossRef]
56. Robles-Ruiz, J.; Baena-Raya, A.; Jiménez-Reyes, P.; Soriano-Maldonado, A.; Rodríguez-Pérez, M.A. Horizontal versus Vertical Force Application: Association with the Change of Direction Performance in Soccer Players. *Eur. J. Sport Sci.* **2023**, *23*, 28–35. [CrossRef]
57. Križaj, J.; Rauter, S.; Vodičar, J.; Hadžić, V.; Šimenko, J. Predictors of Vertical Jumping Capacity in Soccer Players. *Isokinet. Exerc. Sci.* **2019**, *27*, 9–14. [CrossRef]
58. Rouissi, M.; Chtara, M.; Owen, A.; Burnett, A.; Chamari, K. Change of Direction Ability in Young Elite Soccer Players: Determining Factors Vary with Angle Variation. *J. Sports Med. Phys. Fitness* **2017**, *57*, 960–968. [CrossRef]
59. Pimenta, R.; Antunes, H.; Lopes, T.; Veloso, A. Do Repeated Sprints Affect the Biceps Femoris Long Head Architecture in Football Players with and without an Injury History?-A Retrospective Study. *Biology* **2023**, *12*, 96. [CrossRef]
60. Winkler, P.W.; Thorolfsson, B.; Piussi, R.; Snaebjörnsson, T.; Senorski, R.H.; Karlsson, J.; Samuelsson, K.; Senorski, E.H. Sport-Specific Concomitant Injuries, Return-to-Sport Rates and Second Anterior Cruciate Ligament (ACL) Injuries in Adolescents with ACL Reconstruction. *Br. J. Sports Med.* **2025**, 1–10. [CrossRef]

applied sciences

MDPI

Article

Validity of a User-Friendly Spreadsheet Designed for an In-Depth Analysis of Countermovement Bipodal Jumps

Gabriel Delgado-García [1,*], Alejandro Pérez-Castilla [2,3], Francisco Javier Rojas-Ruiz [4], Ismael Navarro-Marchal [4,5] and Anabel Caballero-Villalta [5,6]

1 SER Research Group, Physical Activity and Sport Sciences Department, CESAG, Comillas Pontifical University, 07013 Palma de Mallorca, Spain
2 Department of Education, Faculty of Education Sciences, University of Almería, 04120 Almería, Spain; alexperez@ual.es
3 SPORT Research Group (CTS-1024), CIBIS Research Center, Faculty of Education Sciences, University of Almería, 04120 Almería, Spain
4 Department of Physical Education and Sport, Faculty of Sport Sciences, University of Granada, 18011 Granada, Spain; fjrojas@ugr.es (F.J.R.-R.); ismaelnavarro71@gmail.com (I.N.-M.)
5 Human Lab, Sport and Health University Research Institute (iMUDS), Faculty of Sport Sciences, University of Granada, 18007 Granada, Spain; anabelcaballero@ugr.es
6 Department of Didactics of Musical, Plastic and Corporal Expression, Faculty of Education Sciences, University of Granada, 18011 Granada, Spain
* Correspondence: gadelgado@comillas.edu

Abstract: Countermovement bipodal jumps (CMJs), are widely used for health and performance assessment, but the software required for such analyses is often costly. The study aim was to examine the validity of an Excel/VBA spreadsheet for comprehensive CMJ kinetic analysis. The outcomes have been informed from scientific literature, and the spreadsheet includes modules for data filtering and photogrammetric analysis. To evaluate its validity, 21 participants performed CMJs on a dual force platform system, and the primary outcomes were compared with those derived from prestigious software (MARS 4.0). When jump height was calculated based on take-off speed and flight time the Mean Absolute Errors were 1.79 and 0.69 cm and the minimal detectable changes (MCD) were 1.28 and 0.16 cm. For the propulsive impulse, the error was 5.5 N · s and the MCD was 5.41 N · s. The intraclass correlations were 0.932 (0.902–0.953), 0.984 (0.977–0.989), and 0.940 (0.914–0.959), respectively, demonstrating a strong relationship, and residuals exhibited homoscedasticity. Considering the variability reported in previous studies for intra- and inter-subject comparisons, these errors are minimal, highlighting the spreadsheet's sensitivity. With its exhaustive analytical capabilities and customizable features, this template serves as a valuable tool for trainers, physiotherapists, and academic teaching settings.

Keywords: affordable solution; force plate; fundamental movement assessment; ground reaction force and vertical jump

1. Introduction

Physical performance and health evaluation are crucial in both high-performance sports and clinical practice. Jump tests are widely recognized as key indicators of an individual's functional capacity, explosive strength, and neuromuscular status [1,2]. Prominently, health assessment protocols across diverse populations incorporate jump measurements to assess strength and balance—two fundamental elements in fall prevention and longevity promotion [3]. There is also a relationship between jump height, sprint, and muscular strength, suggesting that jump outcomes could be used as an indicator of sport performance [4]. Furthermore, asymmetries in jump performance, where one limb generates more force than the other, often reflect underlying muscular imbalances or compensatory mechanisms that disproportionately overload one limb, thereby exacerbating the risk of injury in sports and other high-intensity physical activities [5]. These studies emphasize the significance of jump tests as essential tools in both athletic and public health evaluations. An in-depth analysis of the countermovement bipodal jump (CMJ) requires analyzing each leg's variables in isolation, not only taking into account the magnitude of the ground reaction forces but also the temporal offset between each phase of the dominant and non-dominant leg.

In this context, force platforms are frequently regarded as the benchmark instrument for assessing the kinetic and kinematic variables associated with jumping [6,7]. The accuracy and reliability of the data obtained from these platforms are contingent upon the software employed, as the precision of variable calculations hinges on the formulas utilized [8]. A notable example is Measurement, Analysis, and Reporting Software [MARS 4.0], developed by the esteemed brand Kistler. MARS 4.0 is a comprehensive solution that requires no specialized programming knowledge, making it accessible for broad use in both research and practice [9]. This software facilitates the extraction of a wide array of parameters, including kinematical and temporal parameters or kinetics ones [10]. Among these, two of the most widely used in research are jump height calculated via take-off velocity or flight time [8,11–14] and propulsive impulse [15,16], which directly influences take-off velocity and, consequently, jump height.

One of the primary challenges associated with software specifically designed to evaluate jumps is its prohibitive cost, with most brands exceeding one thousand euros. Consequently, many coaches and research laboratories with limited budgets find it unaffordable. This underscores the need to develop cost-effective solutions for assessing bipodal CMJs, given the considerable amount scientific literature dedicated to this jumping skill. Conventional spreadsheet-based software (such as Microsoft Excel) offers an alternative, given its widespread availability and user-friendly interface and has been used for complex sport sciences applications [17,18]. Furthermore, Excel includes features that are particularly advantageous in academic settings, such as customizable formulas that can be easily modified and automated (even including more sophisticated script in VBA computer programming language). Indeed, few authors have developed spreadsheets enabling the straightforward calculation of bilateral CMJ parameters, including eccentric and concentric phase impulses, as well as jump height derived from flight time or take-off velocity [19,20]. However, there is currently no solution available that facilitates the assessment of CMJ using a dual-platform system, allowing for the isolated analysis of each leg. In this context, the existing literature emphasizes that the choice between unilateral or bilateral jumps should be dictated by the specific objectives of the study: unilateral CMJ provides a more accurate indicator of individual limb performance, whereas bilateral CMJ offers a broader understanding of inter-limb compensatory strategies [20,21].

The present study aimed to compare, in the particular case of the CMJ, the primary outcomes (jump height and propulsive impulse) generated by an ad hoc spreadsheet with those derived from prestigious software (MARS 4.0). It is hypothesised that the error would be small and that the spreadsheet is sensitive enough to capture changes in future intra- and inter-subject studies. Given that the spreadsheet allows a good number of variables to be calculated and that it is customizable, its use can be extended to rehabilitation, sports performance, and academic teaching settings.

2. Materials and Methods

This cross-sectional observational study examined the differences between reliable, valid, and user-friendly software (MARS 4.0; Milwaukee, WI, USA) and a freely available, ad hoc Excel/VBA spreadsheet (Excel version: Microsoft® Excel® LTSC MSO [16.0.14332.20824] 64-bit) for the analysis of jumping biomechanics. MARS 4.0 software has demonstrated high intra-subject reliability [9] and has been utilized in studies evaluating the validity of various instruments [22]. Due to its proven reliability and widespread use in the field, MARS 4.0 software could be regarded as a benchmark in biomechanical assessments.

2.1. Sample

This study was conducted with a cohort of 21 healthy and physically active male students from the Faculty of Sports Sciences at the University of Granada, aged between 18 and 25. Participants had to meet the following inclusion criteria: (i) being young, healthy adults aged 18 to 30 years, without musculoskeletal dysfunction. Subjects with any of the following criteria were excluded: (i) surgical interventions in the last six months, (ii) any disability that made it difficult to perform the tests correctly, (iii) any pain that could affect the outcome of the tests, (iv) the technical inability to execute a CMJ jump. None of the participants had experienced any musculoskeletal injury in the three months preceding this study that could potentially impair their jumping ability. Participants were informed of the benefits and risks of the investigation prior to signing an institutionally approved informed consent form. This study adhered to the ethical guidelines outlined in the Declaration of Helsinki for human research and was approved by the Research Ethics Committee of the University of Granada (nº: 687/CEIH/2018).

2.2. Instruments

The tools employed in this study included MARS 4.0 software, an Excel/VBA spreadsheet (including VBA script), and a force platform. Kistler MARS 4.0 (Measurement, Analysis, and Reporting Software) encompasses various kinetic analysis modules, including ones specifically for jump analysis (e.g., CMJ, Squat, Drop Jump, and Hop Jump) [10]. In this study, in addition to the MARS 4.0 analysis itself, the data for each jump were imported into the custom-made Excel/VBA spreadsheet. A selected set of variables was then calculated and compared with the MARS 4.0 outcomes: (i) jump height computed from take-off velocity, (ii) jump height computed from flight time, and (iii) force impulse between maximal force and take-off. This last variable was used as an approximation of the propulsive impulse, since MARS 4.0 does not provide this variable directly, and it was assumed that the propulsive phase begins approximately when the force reaches its maximum positive value. The force platform utilized in this research was a Kistler 9286BA (60 × 40 × 3.5 cm; Winterthur, Switzerland), a portable piezoelectric force platform with a maximum sampling frequency of 1000 Hz, a 16-bit AD converter, and a voltage range of ±10 volts. Kistler is an internationally recognized brand, and its platforms have been employed as benchmark systems in studies validating jump height assessment instruments [8,22].

The following subsection provides a detailed explanation of the use of the ad hoc Excel/VBA spreadsheet, enabling other researchers to effectively implement it. The structure and usage of the spreadsheet will be thoroughly discussed. Additionally, the spreadsheet, accompanied by two video tutorials, some CMJ force data examples (in .txt format), and a pdf document including a list of the variables provided with percentiles (based on the 105 jumps recorded for this study) is available as Supplementary Material.

2.2.1. Structure and Use of the Excel/VBA Spreadsheet

The Excel/VBA tool has several sheets and VBA modules with comments explaining the function of each of these modules. The Excel/VBA tool sheets are as follows: (i) Import data (Sheet 1), were data obtained from a force plate and can be imported (based on VBA open code); (ii) Input (Sheet 2), a sheet that contains the force data from both plates, copied directly from the sheet Import data or pasted by the user; (iii) CMJ analysis (Sheet 3), the sheet containing the formulas and graphs necessary for performing the analysis; (iv) Report (Sheet 4), a sheet that can be printed, and that includes some CMJ outcomes; (v) Percentiles (Sheet 5), for the variables computed by the spreadsheet based on the data of the present manuscript (same as included in Supplementary Material); (vi) Photo analysis (Sheet 6), where a photogrammetric analysis can be performed based on a photo sequence of the jump; (vii) Database (Sheet 7), where the variables extracted from each jump can be stored; and (viii) Help! How to set the landmarks (Sheet 8) and (ix) Help! How to know photo path (Sheet 9), which provide brief instructions for selecting the landmarks (events) required to perform the jump analysis and to select the path of the photo sequence. Sheets 1, 2, and 3 are explained in further detail below, while sheets 4, 5, 7, and 8 do not require user manipulation or are simple to manage. Sheet 6 will be explained in another subsection.

Sheet 1 (Import_data) facilitates the uploading and filtering of data from a .txt file, a common format used by many tools. The sheet is designed to import .txt files from two force data logging software programs, MARS 4.0 and Bioware (V5.4.2.0, Kistler Ibérica SL, Barcelona, Spain). The data are organized according to the .txt file format. For example, the MARS 4.0 software generates a .txt file containing four columns: the time column, which records the time data for each entry; the Fz (N) column, which presents the resultant force values from the two platforms used; the FzLeft (N) column, which displays the force data on the Z-axis of the left-leg force platform; and the FzRight (N) column, which shows the force data on the Z-axis of the right-leg platform. All the forces represent the vertical ground reaction force. If users want to import the data in another format, they have to modify the VBA script (it does not require high level programming knowledge). The Excel/VBA spreadsheet includes a fourth-order zero-phase shift Butterworth low-pass data noise filter with a user-defined cut-off frequency, implemented in VBA by Van Wassenbergh [23] and based on the book of Winter [24].

On Sheet 2 (Input), data from the Import Data sheet is reorganized to derive the kinetic variables of the CMJ. The user must specify the platform's sampling frequency in cell E1 and the gravitational acceleration value in $m \cdot s^{-2}$ in cell E3.

Sheet 3 (CMJ Analysis) is crucial for performing the kinetic analysis of the jump. It relies on Excel formulas and VBA programming. A video demonstrating the tool's usage and sample files with jump kinetic data are provided as supplementary materials. To assist users in understanding these instructions, different cell ranges (matrices) are color-coded, and plots are outlined in various tones (see Figure 1):

- List of matrices: Orange Matrix (A1:C13), Green Matrix (A19:I5020), White Matrix (J19:K5020), Grey Matrix A (L19:Q5020) and Grey Matrix B (AD1:AY11), Blue Matrix A (R19:W5020) and Blue Matrix B (AD67:AY77), and Red Matrix A (X19:AZ5020) and Red Matrix B (AD132:AY142).
- List of plots: Pink Border Plot (in cells range D1:R14), Black Border Plot (in cells range AD21:AW63), Blue Border Plot (in cells range AD87:AW189), and Red Border Plot (in cell range AD151:AW194).

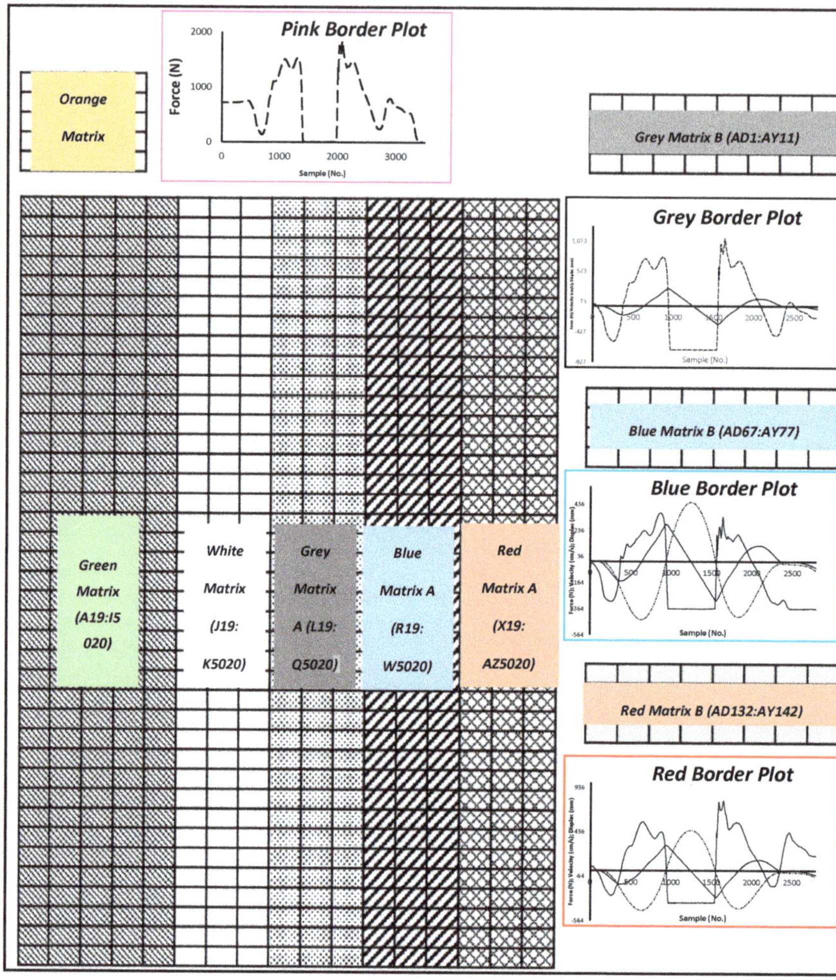

Figure 1. Schematic representation of the Excel sheet.

The first step for the user is to click on the button Manual Calculation (cell B1) to prevent the sheet from performing calculations that could potentially freeze the program when modifying the data in the graphs. After this, in the Pink Border Plot (Figure 2), the following parameters should be determined using the corresponding slider bars located below that plot: (i) the start of the baseline (marker B1), which will facilitate the estimation of body weight; (ii) the end of the baseline (marker B2), which should be approximately two seconds from point B1 (or 2000 samples if the platform records at 1000 Hz); (iii) the start of the jump (marker labelled as Start in the graph), just before the countermovement phase; and (iv) the end of the jump (marker labelled as Finish in the graph), once the force returns to the subject's weight (or a position close to it). The data selected from the graph between the Start and Finish points will be displayed in the Black Border Plot, Blue Border Plot, and Red Border Plot. Once these four points have been selected, the Automatic Calculation Button can then be clicked.

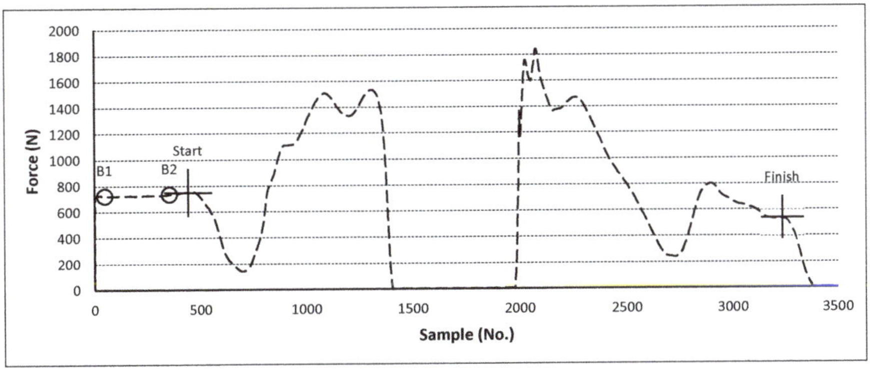

Figure 2. Pink Border Plot, used to select the baseline and the start and finish of the CMJ.

The user of the Excel/VBA spreadsheet will then mark the start and end of the x-axis on the chart (to calibrate the length of that axis and compute a scaling factor in cell BF2 that transforms pixels into graph coordinates) and the ten key points of the CMJ (Figure 3) [19]:

1. Unweighting phase start. The beginning of the CMJ or the start of the negative impulse phase.
2. Braking phase start (or positive impulse start). The end of the negative impulse phase.
3. Propulsive phase start. The end of the countermovement phase, where the velocity equals zero and the negative phase impulse equals the positive phase impulse.
4. Positive impulse finish (the force signal crosses the baseline).
5. Flight phase start. The moment when the take-off begins and the platform does not register any force.
6. Landing time start (or braking landing impulse start). When contact with the platform is re-established.
7. Landing positive impulse start. The start of the positive impulse during landing.
8. Braking landing impulse finish (or repositioning phase start). The end of the braking phase (the velocity equals zero).
9. Landing positive impulse finish (or unweighting phase start). When the force signal crosses the baseline again.
10. Landing unweighting phase end. The end of the CMJ.

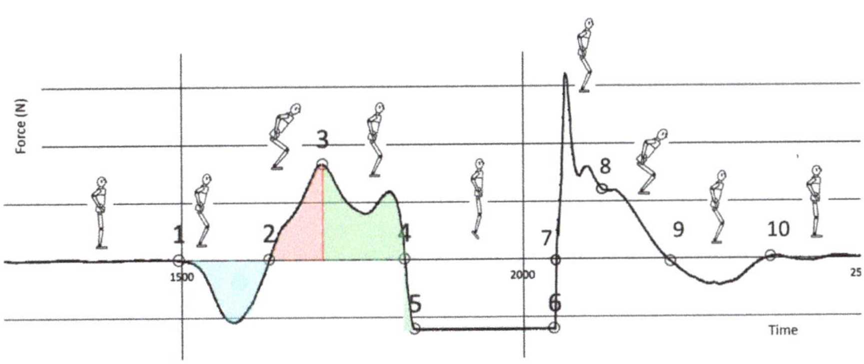

Figure 3. Key event with explanatory illustrations of the CMJ events to be marked on the Grey Border Plot. Color blue represents the unweighting phase impulse, color red represents the braking phase impulse and color green represents the propulsive phase impulse.

To position the key points on the graphs, simply click on the graph at the desired key point location. It is possible to utilize the controls located in cells AW17:AZ17, AW84:AZ84, and AW149:AZ149 to fine-tune the position of the points. In cells AW17, AW84, and AW149 the user selects the point they wish to modify, and in cells AZ17, AZ84, and AZ149, they specify the number of frames the point steps forward or backward on the plot. If there is any previous data, it is possible to delete it by pressing the control near the bottom-right corner, which includes a counter for the number of points selected, ranging from zero to twelve (with two points used for calibration and ten as key points or landmarks).

To check that the key points have been chosen correctly, the user should look at cells AE14, AE17, and AQ15 for the total force, cells AE80, AE83, and AQ82 in the case of the left-leg platform, and cells AE145, AE148, and AQ146 in the case of the right-leg platform. The values in these cells are calculated by subtracting the following: (i) the unweighting phase impulse and the braking phase impulse (cells AE14, AE80, and AE145); (ii) the propulsive positive impulse and the jump break phase impulse (cells AE17, AE83, and AE148); and (iii) the repositioning impulse and the landing unweighting phase impulse (cells AQ15, AQ82, and AQ146). The subtraction of these force impulses must be as close to zero as possible. It will almost never be possible for the value to be zero, due to factors such as sampling frequency (perhaps not sufficient to capture all of the force signal data) or deformation of the shoe sole (causing not all of the force to be applied to accelerate the center of mass). For the selection of the unweighting phase start (event 1) and the take-off time (event 5) the user must also look at cells AJ14, AJ17, AJ80, AJ83, AJ145, and AJ148, where the resultant force (in Newtons) in the unweighting phase start and the difference (in Newtons) between resultant force and half body weight in the flight phase start are displayed. A number of research studies have analyzed the influence of these thresholds on the reliability and magnitude of CMJ outcomes and are recommended to be reviewed for a more precise analysis [25,26].

The pixel data are generated in column BD2:BD13, and the data are then converted into real coordinates in cells BE2:BE13 (based on the scale factor calculated in cell BF2). The user should repeat this procedure to analyze the data for each platform separately in the Blue Border Plot and the Red Border Plot (to analyze the phases of each leg). If the user wants to perform the analysis based on global force data (without considering the data from each platform in isolation), they should paste the data into cells BI and BN. This approach may be useful for studying the contribution of each leg during each phase of the CMJ. If the user wishes to examine the time lag between the two legs, it would be more convenient to perform the analysis separately for each platform.

The computation of both continuous signals and discrete variables is detailed in the following two subsections.

2.2.2. Continuous Kinetic Variables Computation Formulas

The Orange Matrix contains a series of data necessary for calculating the kinetic variables: (i) in cell B4, the acceleration of gravity; (ii) in cells B5 and B6, the time at the start and end of the baseline measurement; (iii) in cells C5 and C6, the force at those moments; (iv) in cells B7 and B8 and C7 and C8, the time at the start and end of the jump and the corresponding force values; and (v) in cells B9 and B13, a series of variables related to body weight:

- Cell B9. The weight in Newtons is calculated based on the values between points B1 and B2 (baseline) on the pink-bordered graph.
- Cell B10. The body mass in kg is determined from the weight (cell B9) and the acceleration of gravity (cell B4).

- Cells B11 and B12 display the weight recorded with an external scale, indicated on the Input sheet, which serves as a confirmatory measure to ensure consistency with the calculated weight value.
- Cell B13 contains half of the body weight.

To calculate the kinetic variables of the CMJ, velocity and displacement at each instant are determined through an integration process based on instantaneous acceleration. This process is explained in detail below. The sheet is designed to accommodate up to 5000 records. If the platform records at 1,000 Hz, this corresponds to a recording duration of 5 s (with each row representing one millisecond). The sampling number is displayed in the Green Matrix, column A. Columns B, C, E, and F contain data from the "Input" sheet. Columns D, F, and H calculate the resultant force: column D shows the resultant force by subtracting the force recorded by the platform from the participant's weight (column D) or half of the body weight (columns F and H). The White Matrix displays the rows containing the jump data, as selected on the Pink Border Plot. The Grey Matrix shows data from the resultant force (limited to data selected in the Pink Border Plot), the Blue Matrix displays data from the force platform under the left leg, and the Red Matrix includes data from the force platform under the right leg.

Columns L, R, and X display the instantaneous total resultant force, the force recorded on the left platform and the force on the right platform, respectively. Columns M, S, and Y contain the instantaneous acceleration data, calculated by dividing the instantaneous force by the weight. Columns N, T, and Z present the instantaneous velocity, derived from the instantaneous acceleration and the velocity of the preceding row. In columns O, U, and AA, displacement is calculated using the instantaneous velocity and the displacement from the previous row. The instantaneous power (columns P, V, and AB) is determined based on the instantaneous force and velocity. Finally, the instantaneous impulse (columns Q, W, and AC) is computed using the instantaneous force and the sampling frequency. All these calculations are summarized in Table 1.

Table 1. Computation of the continuous kinetic variables.

Matrix	Variable	Excel Col.	*Computation*
Grey	Net resultant force	D, L	Subtracting the participant's weight to the resultant force at each instant
	Acceleration of the GC	M	By dividing the net resultant force at each instant by the weight of the participant
	Velocity of the GC	N	Adding to the instantaneous velocity at previous record, the instantaneous acceleration multiplied by the time of one record
	Displacement of the CG	O	Adding to the instantaneous displacement at previous record, the instantaneous velocity multiplied by the time of one record
	Instantaneous power of the CG	P	Multiplying the net resultant force at each instant by the velocity of the CG at each instant
	Instantaneous impulse of the CG	Q	Multiplying the net resultant force at each instant by the time of one record

Table 1. *Cont.*

Matrix	Variable	Excel Col.	*Computation*
Blue A and Red A	Net resultant force of the left/right body side	F, H, R, X	Subtracting half of the body weight to the left or right force at each instant
	Left or right body side acceleration	S, Y	By dividing the net left/right force at each instant by half of the body weight
	Instantaneous velocity of the left or right body side	T, Z	Adding to the instantaneous velocity of the left or right body side in the previous record, the instantaneous acceleration of the left or right body side multiplied by the time of one record
	Displacement of the left or right body side	U, AA	Adding to the instantaneous displacement in the previous record, the instantaneous velocity of the left or right body side multiplied by the time of one record
	Instantaneous power of the left or right body side	V, AB	Multiplying the net resultant force of the left or right body side at each instant by the velocity of the left or right body side at each instant
	Instantaneous impulse of the left or right body side	W, AC	Multiplying the net resultant force of the left or right body side at each instant by the time of one record

2.2.3. Discrete Kinetic Variables Computation Formulas

In the Grey Matrix B, several variables related to the quality of the jump are calculated, taking into account the resultant force. Columns AE and AF list the names and numbers of each key event of the jump. The AG column indicates the row number or sample in which the event appears, while the AH column specifies the second in which the event occurs. Further important discrete variables are detailed in Table 2.

Table 2. Computation of the discrete kinetic variables.

Matrix	Variable	Excel Column	*Computation*
Grey B, Blue B, and Red B	Partial impulses	AJ	Sum of the instantaneous impulses between the two selected events
	Propulsive average velocity and braking average velocity	AL	Averaging the velocity values between the start of the propulsive phase and the start of the flight phase and between the start of landing and the end of the braking phase
	Average Force	AN	Dividing the impulse of each phase by the phase duration
	Average Power	AO	Multiplying the average force in each phase by the average velocity in each phase

The total analysis time of a CMJ using this tool is under five minutes, as demonstrated in the Supplementary Video that explains the use of this spreadsheet. The Excel spreadsheet produces a report table containing several computed parameters (Table 3) included in the MARS 4.0 software, many of them are extensively used in the literature [27].

Table 3. Report generated by the Excel/VBA spreadsheet (example of a CMJ).

General Measures		Phase Delays (ms)	
Excentric impulse (N · s)	113.5	Unweighting phase start	−51
Propulsive impulse (N · s)	194.0	Braking phase start delay	35
Impulse ratio	0.585	Positive impulse finish	−5
Take-off speed (m/s)	2.62	Flight phase start	−5
Jump height with initial velocity (cm)	35.14	Landing time start delay	−5
Flight time (Sec)	0.57	Landing positive impulse start delay	0
Jump height with flight time (m)	39.94	Braking landing impulse finish delay	−289
Propulsive relative power (W/kg)	14.8	Landing positive impulse finish delay	−282
Landing positive impulse (N · s)	308.6	Landing unweighting phase end delay	−15
Landing peak force (N)	1117.9	**Phase duration differences (ms)**	
Left-leg negative impulse (N · s)	−61.6	Unweighting phase start	86
Right-leg negative impulse (N · s)	−47.2	Braking phase start delay	−30
Inter-leg negative impulse asymmetry (R-L) (%)	26.5	Positive impulse finish	0
Left-leg positive impulse (N · s)	106.6	Flight phase start	0
Right-leg positive impulse (N · s)	200.2	Landing time start delay	5
Inter-leg positive impulse asymmetry (R-L) (%)	−61.0	Landing positive impulse start delay	−289
Left-leg landing positive impulse (N · s)	58.76	Braking landing impulse finish delay	7
Right-leg landing positive impulse (N · s)	230.7	Landing positive impulse finish delay	267
Inter-leg landing PI asymmetry (R-L) (%)	−118.8	Landing unweighting phase end delay	0
Landing force peak Left leg (R-L) (N)	343.6		
Landing force peak Right leg (R-L) (N)	791.0		
Landing force peak asymmetry (R-L) (%)	−78.9		

2.2.4. Performing a Photogrammetric Analysis

As mentioned above, the Excel/VBA template allows for an additional photogrammetric analysis in Sheet 6, Photo analysis. This can be useful to check that the CMJ has been performed correctly, to provide feedback to the participant being evaluated, to check that the events have been marked correctly, or for educational purposes, to show the start and end frame of each phase.

This analysis requires recording a photo sequence of the CMJ, which must be saved in the computer folder where the Excel file is located. Each photo should be numbered as Photo (1), Photo (2), etc. This is easily performed in the Window environment by selecting all the photos and naming the first one as Photo (1). Once this is performed, the user should modify cells C1 (selecting the variable number from the range A3:A7, which will be synchronized with the photo frame in the graph on the right), I1 (it is recommended to see Sheet 9 Help! How to know the path of the photo), I2, I4, and Q3 (selecting the frame of the associated folder where the take-off of the CMJ takes place). Next, the user must modify the maximum value of scroll bar 1 by right-clicking on it and setting the maximum value equal to the number of photos in the associated folder. Finally, it is possible to navigate between frames by clicking on the rotate button or the scroll button. In the graph on the right there is a red dot corresponding to the CMJ photo frame (Figure 4).

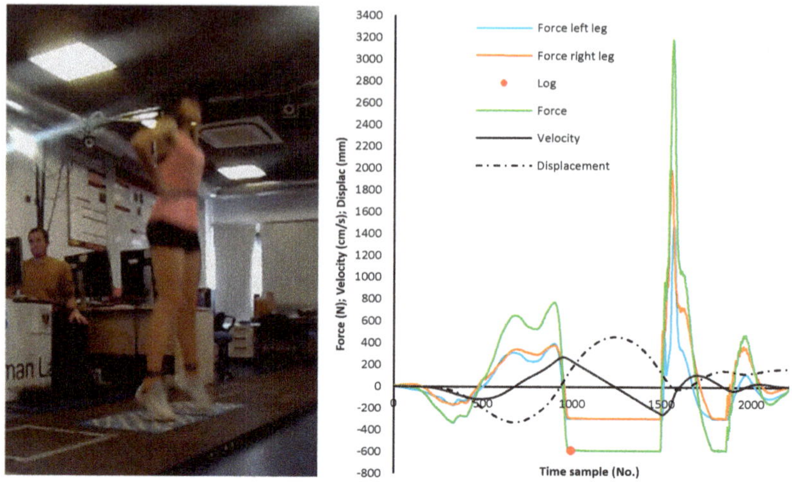

Figure 4. Photo sequence and force data synchronized in Sheet 6 (Photo analysis).

2.3. Procedures to Analyze the Validity of the Spreadsheet

After a standardized warm-up—comprising five minutes of continuous running, a protocol of five minutes of dynamic stretching targeting the muscles most engaged during the jumps, and five CMJs—the participants undertook the jumping protocol. The data were recorded using MARS 4.0 software, which provides several variables semi-automatically. The raw force data was also imported into the ad hoc Excel/VBA spreadsheet, filtered with a cut-off frequency of 100 Hz [28] and analyzed. Jump height, determined from take-off velocity or flight time, and propulsive impulse outcomes were compared between MARS 4.0 and the ad hoc spreadsheet.

Prior to the jump, the following instruction was given: "Jump as high as you can, initiating the jump from an upright stance and with hands on the hips and performing a countermovement". The CMJ should be performed involving a rapid downward movement followed by an upward movement (countermovement) to utilize the elastic component of the musculature for achieving a greater jump height. The participants were instructed to hold a static position for approximately two seconds before the jump to establish a baseline for calculating body weight. Each participant performed five jumps, with a two-minute rest between each. If the subject mistakenly performed arm swings during the jump, it was repeated.

2.4. Statistical Analysis

All the statistical analyses were conducted using Microsoft Excel 2018. All the data were tested for normality with a visual inspection of the QQ-plots and of the histograms. To compare the results from the Excel/VBA spreadsheet with those from the MARS 4.0 software, the following statistical issues were analyzed for each of the variables: (i) the residual homoscedasticity via Bland–Altman plots including the regression lines for the residuals; (ii) the magnitude of the error in both absolute and relative terms with the Mean Absolute Errors (MAE), Standard Error of Measurement (SEM), and Percentage Mean Absolute Error relative to the average measurement obtained with the MARS 4.0 software (PMAE); (iii) the level of agreement based on the Lin's intraclass correlation coefficient (Lin CCC) with the 95% confidence intervals. The strength of agreement for the Lin CCC was categorized as follows: > 0.8, very strong; 0.6–0.8, strong; 0.4–0.6, medium; 0.2–0.4, weak; <0.2, very weak [29]. Finally, to determine the magnitude of change that would exceed the threshold of measurement error [30], the minimal detectable change (MDC) as a 90% and

95% confidence interval (MCD_{90} and MCD_{95}) was calculated. It was performed with the following formulas: $MCD_{90} = SEM \cdot \sqrt{2} \cdot 1.65$ and $MCD_{95} = SEM \cdot \sqrt{2} \cdot 1.96$.

3. Results

The means and standard deviations of the height obtained with MARS 4.0 and with the ad hoc Excel/VBA spreadsheet were 36.5 ± 7.3 cm and 37.2 ± 7.2 cm when calculated using take-off velocity and 36.5 ± 6.6 cm and 35.8 ± 6.6 cm when calculated using flight time. The mean propulsive impulse was 207 ± 20 N · s and 209 ± 20 N · s.

For jump height calculated based on take-off velocity, the MAE was 1.79 cm (PMAE was 4.92%). In contrast, for jump height calculated based on flight time, the MAE and PMAE were lower, 0.69 cm and 1.89%, respectively. For the propulsive impulse, the MAE was 5.5 N · s (PMAE = 2.66%). All Lin's CCC values indicated a very strong relationship. The statistical parameters, including the 95% confidence interval for the Lin CCC, the SEM, and the minimal detectable changes, are included in Table 4.

Table 4. Statistical parameters used to evaluate the spreadsheet's validity.

Variable	Statistical Parameter				
	Reliability	Error (cm)		* MCD (cm)	
	Lin CCC (95% CI)	MAE	SEM	MDC_{90}	MDC_{95}
Jump height from take-off velocity	0.932 (0.902–0.953)	1.79	0.55	1.28	1.52
Jump height from flight time	0.984 (0.977–0.989)	0.69	0.07	0.16	0.19
Propulsive impulse	0.940 (0.914–0.959)	5.5	1.95	4.55	5.41

* Lin's concordance correlation coefficient with the lower and upper ends of the 95% confidence interval; * MDC: minimal detectable change; MAE: Mean Absolute Error; SEM: Systematic Error of Measurement.

Residuals depicted in the Bland–Altman plots appeared homoscedastic for all three variables (and the associated regression lines did not exhibit any linear relationship), indicating that the differences between MARS 4.0 and the ad hoc Excel/VBA spreadsheet were not dependent on the magnitude of the measurement (see Figures 5–7).

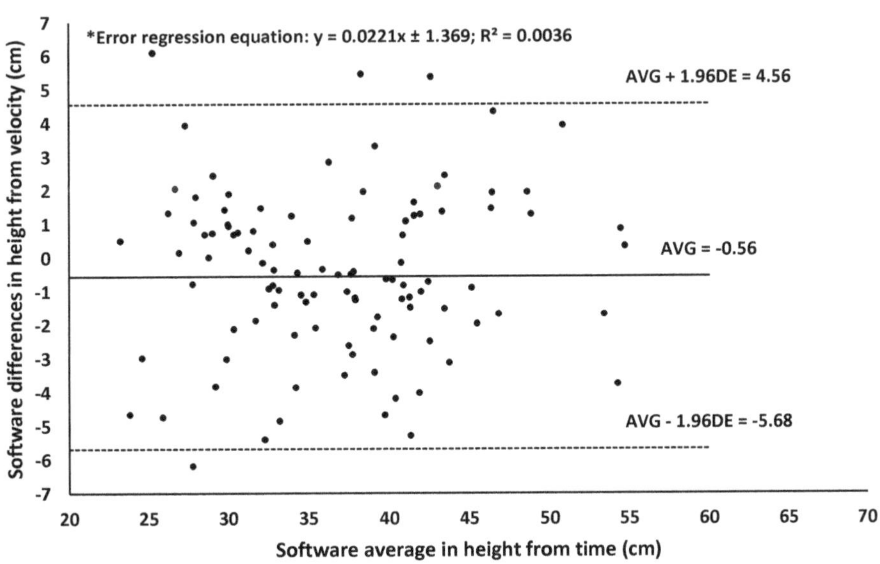

Figure 5. Bland–Altman plot of the height computed from take-off velocity.

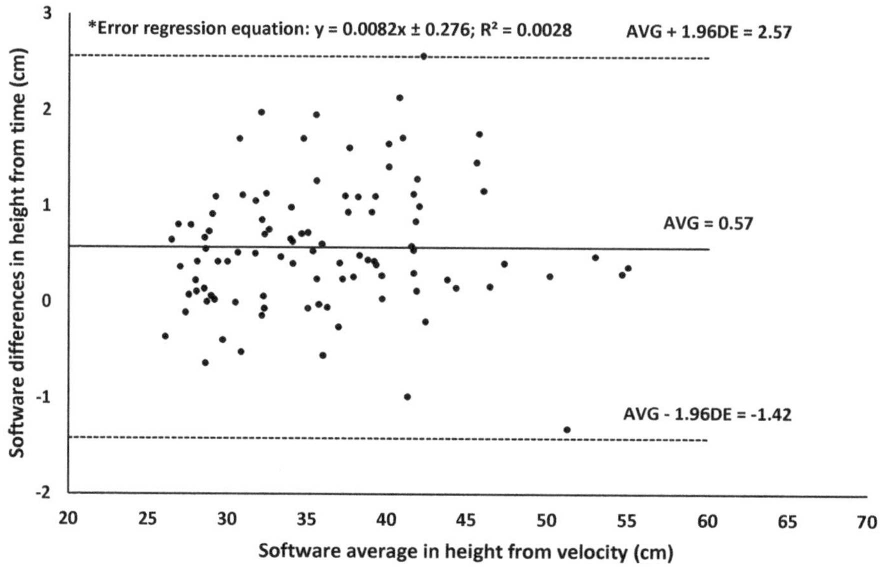

Figure 6. Bland–Altman plot of the height computed from flight time.

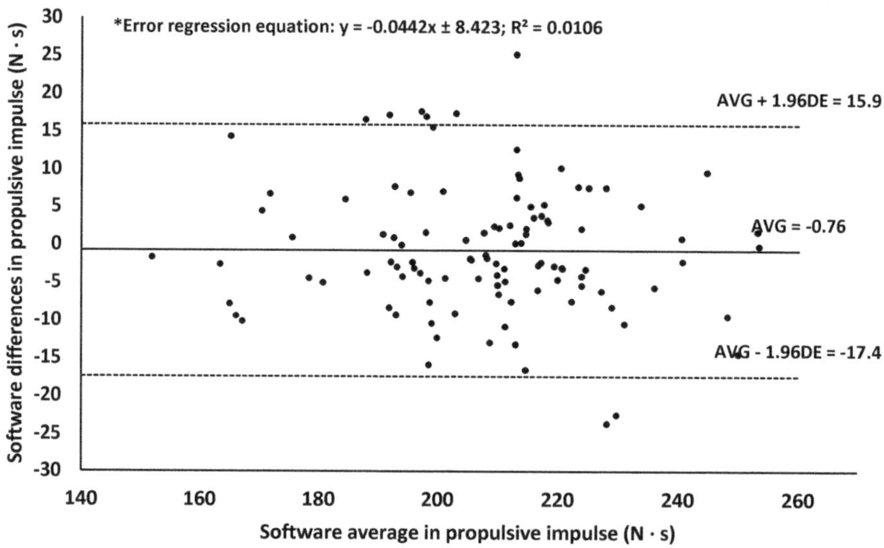

Figure 7. Bland–Altman plot of the propulsive impulse.

4. Discussion

This study introduces a custom Excel/VBA spreadsheet designed specifically for the analysis of dual platform countermovement bipodal jumps, offering a cost-effective and accessible alternative to high-end software solutions. The manual identification of key events during the jump, which can be completed in less than five minutes, allows for the extraction of a wide range of biomechanical variables. These variables are summarized in a printable format, integrating data from both force platforms while also allowing for the individual analysis of each leg's function during the jump. The comparison between the ad hoc Excel/VBA spreadsheet and the prestigious software (MARS 4.0), recognized as a benchmark in biomechanics laboratories, revealed small differences. The difference in jump height estimated using flight time was less than one centimeter, and less than two centimeters when estimated using take-off velocity, corresponding to relative errors of approximately 2% and 4%, respectively. The propulsive impulse showed a difference of approximately 5 N · s, with a relative error of about 2.5%. These findings suggested that the custom spreadsheet provided accurate results, making it a viable option for laboratories and researchers with limited resources.

When assessing the magnitude of these errors, it is essential to consider the typical between-group and within-subject differences reported in the literature. In other words, coaches and scientists must avoid measurement errors larger than the differences they intend to measure [31,32]. For example, a researcher intending to use the ad hoc Excel/VBA spreadsheet for a specific aim should carefully consider the magnitude of error relative to the differences found in similar studies. Regarding jump height, a difference of approximately 1.8 cm was observed when using take-off velocity for estimation and of 0.7 cm when using flight time. Some scientific literature results are discussed below, in which the use of the ad hoc Excel/VBA spreadsheet would be recommended or not. In a study by Markström and Olsson [11] comparing jump performance between throwers and sprinters, the sprinters exhibited a jump height approximately 10 cm greater (35 cm vs. 45 cm). Similarly, in a study by Mihalik et al. [13], a four-week training program incorporating resistance and plyometric exercises, with a frequency of two days per week, demonstrated improvements in jump height ranging from 3 to 5 cm after three weeks. Both examples show differences larger than the spreadsheet errors of 1.8 cm when using take-off speed to estimate jump height and 0.7 cm when using flight time, indicating that it is sensitive enough to accurately detect these variations. In addition, a shoe with higher bending stiffness improved jump height by an average of 1.7 cm [14]. While this improvement exceeds the 0.7 cm error associated with the Excel/VBA spreadsheet using flight time, it is very close to the 1.8 cm error observed when using take-off speed, suggesting that the latter method may not be suitable for such studies. Moreover, Markovic [33] assessed the effects of plyometric and concentric exercise programs lasting between 6 and 12 weeks, finding jump height improvements of approximately 5% to 10% (equating to 2–6 cm). These percentage increases are substantially higher than the percentage error for jump height obtained with the Excel/VBA spreadsheet using flight time, confirming its adequacy for such assessments. However, when take-off speed is used, the percentage error (≈4%) is lower than that reported in most studies included in the meta-analysis but is comparable to those with the smallest improvements. In these latter cases, the use of take-off speed for estimation is not recommended. In a study examining within-subjects jump height changes across sessions, systematic biases of 1.3 cm were found [12]. These biases exceed the Standard Error of Measurement (SEM) of the ad hoc Excel/VBA spreadsheet, whether using take-off velocity or flight time to estimate jump height. The MCD_{90} and MCD_{95} confirm that the spreadsheet is capable of detecting most of the changes discussed, as it can detect changes in jump height of 1.28 or 1.52 cm and 0.16 or 0.19 cm when using the take-off velocity or the flight time, respectively.

Regarding propulsive impulse, fewer studies have focused on this variable, possibly due to the relative ease and interpretability of measuring jump height. Since the MARS 4.0 software does not provide a propulsive impulse measurement directly, the force impulse from maximal force to take-off was used as an approximation. Given that propulsive impulse directly determines take-off velocity and consequently flight height, it is arguably a more direct and potentially less error-prone variable than flight time [34]. An observational study reported propulsive-concentric impulses of 171.18 ± 38.6 N · s in females and 254.1 ± 37.46 N · s in males [35], so the ad hoc Excel/VBA spreadsheet could detect these differences, because it showed a margin of error of 5 N · s. In other related research, the platform recorded concentric impulse values of approximately 230 ± 50 N · s in CMJs [16]. This standard deviation (50 N · s), encompassing both random and systematic errors, is significantly higher than the Excel/VBA spreadsheet's 5 N · s error, indicating that this can be considered a minimal error. Furthermore, the use of arm swing on concentric impulse allows for a gain in propulsive impulse of about 30 N · s [15], a difference also detectable by the Excel/VBA spreadsheet. In this case, the MDC_{90} and the MDC_{95} indicated that a change greater than 4.55 or 5.41 N · s would be required to be certain that the change is not due to measurement error when using the spreadsheet. These values are also below the differences in the means of the studies used as examples, confirming the sensitivity of the measurement.

The present study was not exempt from limitations. Firstly, this study's focus on young, healthy male participants limits its applicability to other groups. Although the Bland–Altman plots did not show any kind of tendency on the residuals, it would be necessary to study whether the template is still accurate with lower jump heights, achieved, for example, by children or sedentary people or even with athletic jumpers reaching greater jump heights. Although no statistical power study was conducted, the sample size was similar to that of studies analysing jumping assessment instruments or procedures [8,25,26,28]. In addition, the events are marked based on visual observation of the signal and there exists a human error that has to be considered. Manual marking of CMJ phases could introduce subjectivity and inconsistencies, particularly for inexperienced users. Such error can be assessed by comparing the marking of events with a fully automatic system and also by performing intra- and inter-rater reliability studies. Nevertheless, the sheet includes calculations that attempt to minimize such error. In particular, three subtractions of force impulses are calculated which should be the closest possible to zero, explained at the end of the Section 2.2.1. On the other hand, considering its wide use in research, only one brand and one force platform software (Kistler and MARS 4.0) was used, and it would be interesting to analyse whether other systems could provide similar accurate results. Finally, a fourth-order, zero-phase-shift Butterworth low-pass filter was selected, based on a previous study [28], but perhaps other types of filters would have been more appropriate (e.g., an adaptive filter [23] or a residual-based filter).

Despite these limitations a notable advantage of the custom Excel/VBA spreadsheet is its adaptability and ease of use. Unlike more complex mathematical software (e.g., MATLAB or Python), it does not require extensive programming expertise, making it accessible to a broader range of users, including those in educational settings (i.e., in university degrees such Sports Science, Kinesiology, or Physiotherapy). The flexibility of the spreadsheet allows researchers to tailor the analysis to specific research questions or educational objectives, enhancing its utility in both research and teaching environments. This is particularly valuable in settings where the budget for specialized software is limited, yet there is a need for reliable and valid biomechanical analysis tools. Furthermore, the spreadsheet's ability to analyze each leg separately during bipodal jumps offers insights into inter-limb coordination and asymmetries. This feature is crucial for identifying compensatory strategies or imbalances that could predispose athletes to injury [35,36]. While the current study focused on jump height and propulsive impulse, future research could expand the validation of this tool to include other kinetic variables, such as rate of force development, explore its reliability across different populations and conditions, or compare the accuracy with that of systems such as 2D and 3D photogrammetry or inertial sensors. The spreadsheet could also provide several physical outcomes useful for training or rehabilitation purposes such as (i) muscle power, (ii) rate of force development, (iii) impact peaks upon landing, (iv) outcomes related to the degree of muscle recovery, or (v) the improvements in muscle strength or jumping ability following an intervention program. It can also be utilized to assess specific research issues such as (i) the effect of different plyometric jump programs on asymmetries [37]; (ii) the inter-limb coordination during jumping, which can also be used as a measure of neuromuscular fatigue e.g., generated by competition [38]; (iii) the effects of shoe soles designed to minimize energy dissipation and improve the jumping performance [14]; or the relation between inter-limb asymmetries and jumping performance [39]. Additionally, inter-rater and intra-rater reliability studies could further establish the robustness of this tool, ensuring its broader applicability in diverse research contexts. Moreover, it is noteworthy that, in the past, the high cost of force plates often restricted their accessibility, compelling modest laboratories to rely on photogrammetric systems or contact mats for jump evaluations. However, with the recent availability of force plates priced under USD 1,000, the utilization of conventional spreadsheets becomes increasingly relevant.

5. Conclusions

This study presented a user-friendly and customizable Excel/VBA spreadsheet designed for detailed jump kinetic analysis using a dual force platform system. The results demonstrated that the spreadsheet produced small errors in calculating jump height—whether based on flight time or take-off velocity—and in estimating propulsive impulse. These small errors, coupled with high intraclass correlation coefficients and homoscedastic residuals, suggest that the tool is adequate for a number of practical applications.

While further validation is required, the custom spreadsheet offers a robust, cost-effective alternative to commercial software for analyzing dual-platform, bipodal jumps. Its accessibility, adaptability, and precision make it a valuable resource for trainers, researchers, and educators.

Supplementary Materials: The following supporting information can be downloaded at: https://www.mdpi.com/article/10.3390/app15031519/s1, Table S1: List of variables with percentiles for young sportsmen; Video S1: Excel/VBA spreadsheet for CMJ analysis and report video tutorial; CMJ force data example 1-MARS 4.0; CMJ force data example 2-bioware; Photo_sequence_CMJ force data example 2-bioware; Excel/VBA spreadsheet for CMJ analysis and report.xlsm.

Author Contributions: Conceptualization, G.D.-G. and A.P.-C.; data curation, G.D.-G.; formal analysis, G.D.-G. and A.P.-C.; funding acquisition, A.P.-C. and F.J.R.-R.; investigation, G.D.-G., I.N.-M., and A.C.-V.; methodology, G.D.-G. and F.J.R.-R.; project administration, A.P.-C. and F.J.R.-R.; resources, G.D.-G., A.P.-C. and F.J.R.-R.; software, G.D.-G.; supervision, A.P.-C. and F.J.R.-R.; validation, G.D.-G., I.N.-M. and A.C.-V.; writing—original draft, G.D.-G. and A.C.-V.; writing—review and editing, G.D.-G., A.P.-C. and A.C.-V. All authors have read and agreed to the published version of the manuscript.

Funding: This study was supported by the Spanish Ministry of Education (FPU22/04211) by the project "SensorSportLab III" (High Council of Sports and Spanish Ministry of Culture and Sport) and by the Ministry of Science and Innovation (PID2023-151166OB-I00SRA) (State Research Agency).

Institutional Review Board Statement: This study was conducted in accordance with the Declaration of Helsinki and approved by the Ethics Committee of The University of Granada (protocol code 430/CEIH/2017, approved the 15th July 2018) for studies involving humans.

Informed Consent Statement: Informed consent was obtained from all subjects involved in this study.

Data Availability Statement: Data are unavailable due to privacy or ethical restrictions. If any researcher is interested in the raw data of this study, please contact the corresponding author by mail.

Acknowledgments: The authors would like to thank all those subjects who participated in this research.

Conflicts of Interest: The authors declare no conflicts of interest.

Abbreviations

The following abbreviations are used in this manuscript:

CMJ	Counter movement jump
BA	Bland–Altman
VBA	Visual Basic for Applications programming language

References

1. Cormie, P.; McBride, J.M.; McCaulley, G.O. Power-Time, Force-Time, and Velocity-Time Curve Analysis during the Jump Squat: Impact of Load. *J. Appl. Biomech.* **2008**, *24*, 112–120. [CrossRef] [PubMed]
2. García-Pinillos, F.; Ramírez-Campillo, R.; Boullosa, D.; Jiménez-Reyes, P.; Latorre-Román, P. Vertical Jumping as a Monitoring Tool in Endurance Runners: A Brief Review. *J. Hum. Kinet.* **2021**, *80*, 297–308. [CrossRef] [PubMed]
3. Ruiz, J.R.; Castro-Piñero, J.; España-Romero, V.; Artero, E.G.; Ortega, F.B.; Cuenca, M.M.; Jimenez-Pavón, D.; Chillón, P.; Girela-Rejón, M.J.; Mora, J.; et al. Field-Based Fitness Assessment in Young People: The ALPHA Health-Related Fitness Test Battery for Children and Adolescents. *Br. J. Sports Med.* **2010**, *45*, 518–524. [CrossRef] [PubMed]
4. Wisløff, U.; Castagna, C.; Helgerud, J.; Jones, R.; Hoff, J. Strong Correlation of Maximal Squat Strength with Sprint Performance and Vertical Jump Height in Elite Soccer Players. *Br. J. Sports Med.* **2004**, *38*, 285–288. [CrossRef] [PubMed]
5. Bishop, C.; Read, P.; Lake, J.; Chavda, S.; Turner, A. Interlimb Asymmetries: Understanding How to Calculate Differences From Bilateral and Unilateral Tests. *Strength Cond. J.* **2018**, *40*, 1–6. [CrossRef]
6. Bishop, C.; Turner, A.; Read, P. Effects of Inter-Limb Asymmetries on Physical and Sports Performance: A Systematic Review. *J. Sports Sci.* **2017**, *36*, 1135–1144. [CrossRef] [PubMed]
7. Šarabon, N.; Kozinc, Ž.; Marković, G. Force–Velocity Profile during Vertical Jump Cannot Be Assessed Using Only Bodyweight Jump and Isometric Maximal Voluntary Contraction Tasks. *Sci. Rep.* **2020**, *10*, 19127. [CrossRef] [PubMed]
8. Attia, A.; Dhahbi, W.; Chaouachi, A.; Padulo, J.; Wong, D.; Chamari, K. Measurement Errors When Estimating the Vertical Jump Height with Flight Time Using Photocell Devices: The Example of Optojump. *Biol. Sport* **2017**, *34*, 63–70. [CrossRef] [PubMed]
9. Hébert-Losier, K.; Martyn Beaven, C. The MARS for Squat, Countermovement, and Standing Long Jump Performance Analyses. *J. Strength Cond. Res.* **2014**, *28*, 1849–1857. [CrossRef]
10. Measurement, Analysis and Reporting Software MARS. Available online: http://mars.s2p.si/ (accessed on 11 October 2024).
11. Markström, J.L.; Olsson, C.-J. Countermovement Jump Peak Force Relative to Body Weight and Jump Height as Predictors for Sprint Running Performances. *J. Strength Cond. Res.* **2013**, *27*, 944–953. [CrossRef]
12. Moir, G.; Shastri, P.; Connaboy, C. Intersession Reliability of Vertical Jump Height in Women and Men. *J. Strength Cond. Res.* **2008**, *22*, 1779–1784. [CrossRef]
13. Mihalik, J.P.; Libby, J.J.; Battaglini, C.L.; McMurray, R.G. Comparing Short-Term Complex and Compound Training Programs on Vertical Jump Height and Power Output. *J. Strength Cond. Res.* **2008**, *22*, 47–53. [CrossRef]
14. Stefanyshyn, D.J.; Nigg, B.M. Influence of Midsole Bending Stiffness on Joint Energy and Jump Height Performance. *Med. Sci. Sports Exerc.* **2000**, *32*, 471. [CrossRef]
15. Heishman, A.D.; Daub, B.D.; Miller, R.M.; Freitas, E.D.S.; Frantz, B.A.; Bemben, M.G. Countermovement Jump Reliability Performed With and Without an Arm Swing in NCAA Division 1 Intercollegiate Basketball Players. *J. Strength Cond. Res.* **2020**, *34*, 546–558. [CrossRef] [PubMed]
16. Merrigan, J.J.; Strang, A.; Eckerle, J.; Mackowski, N.; Hierholzer, K.; Ray, N.T.; Smith, R.; Hagen, J.A.; Briggs, R.A. Countermovement Jump Force-Time Curve Analyses: Reliability and Comparability Across Force Plate Systems. *J. Strength Cond. Res.* **2023**, *38*, 30–37. [CrossRef] [PubMed]
17. Delgado-García, G.; Vanrenterghem, J.; Mildenberger, C.; Gallego, L.R.; Chicano-Gutiérrez, J.M.; Manuel Soto-Hermoso, V. Accuracy and Reliability of a Low-Cost Methodology to Assess 3D Body Posture Based on Commercial Cameras and Excel Templates. *Measurement* **2021**, *173*, 108638. [CrossRef]
18. Chavda, S.; Bromley, T.; Jarvis, P.; Williams, S.; Bishop, C.; Turner, A.N.; Lake, J.P.; Mundy, P.D. Force-Time Characteristics of the Countermovement Jump: Analyzing the Curve in Excel. *Strength Cond. J.* **2018**, *40*, 67–77. [CrossRef]
19. McMahon, J.J.; Suchomel, T.J.; Lake, J.P.; Comfort, P. Understanding the Key Phases of the Countermovement Jump Force-Time Curve. *Strength Cond. J.* **2018**, *40*, 96–106. [CrossRef]
20. Cohen, D.; Burton, A.; Wells, C.; Taberner, M.; Diaz, M.A.; Graham-Smith, P. Single vs. Double Leg Countermovement Tests: Not Half an Apple! *Aspetar Sport. Med. J.* **2020**, *9*, 34–41.

21. Pérez-Castilla, A.; García-Ramos, A.; Janicijevic, D.; Delgado-García, G.; De la Cruz, J.C.; Rojas, F.J.; Cepero, M. Between-Session Reliability of Performance and Asymmetry Variables Obtained during Unilateral and Bilateral Countermovement Jumps in Basketball Players. *PLoS ONE* **2021**, *16*, e0255458. [CrossRef]
22. Koltermann, J.; Gerber, M.; Beck, H.; Beck, M. Validation of the HUMAC Balance System in Comparison with Conventional Force Plates. *Technologies* **2017**, *5*, 44. [CrossRef]
23. Van Wassenbergh, S. VBA Excel Add-In Utilities and Worksheet Templates. Available online: https://www.uantwerpen.be/en/staff/sam-vanwassenbergh/my-website/excel-vba-tools/ (accessed on 11 October 2024).
24. Winter, D.A. *Biomechanics and Motor Control of Human Movement*; John Wiley & Sons: Hoboken, NJ, USA, 2009; ISBN 9780470398180.
25. Pérez-Castilla, A.; Fernandes, J.F.T.; Rojas, F.J.; García-Ramos, A. Reliability and Magnitude of Countermovement Jump Performance Variables: Influence of the Take-off Threshold. *Meas. Phys. Educ. Exerc. Sci.* **2021**, *25*, 227–235. [CrossRef]
26. Pérez-Castilla, A.; Rojas, F.J.; García-Ramos, A. Reliability and Magnitude of Loaded Countermovement Jump Performance Variables: A Technical Examination of the Jump Threshold Initiation. *Sports Biomech.* **2019**, *21*, 622–636. [CrossRef]
27. Bishop, C.; Jordan, M.; Torres-Ronda, L.; Loturco, I.; Harry, J.; Virgile, A.; Mundy, P.; Turner, A.; Comfort, P. Selecting Metrics That Matter: Comparing the Use of the Countermovement Jump for Performance Profiling, Neuromuscular Fatigue Monitoring, and Injury Rehabilitation Testing. *Strength Cond. J.* **2023**, *45*, 545–553. [CrossRef]
28. Ferro, A.; Floría, P.; Villacieros, J.; Muñoz-López, A. Maximum Velocity during Loaded Countermovement Jumps Obtained with an Accelerometer, Linear Encoder and Force Platform: A Comparison of Technologies. *J. Biomech.* **2019**, *95*, 109281. [CrossRef] [PubMed]
29. Campbell, M.J.; Jacques, R.M. *Statistics at Square Two*; John Wiley & Sons: Hoboken, NJ, USA, 2023; ISBN 9781119401391.
30. Salamh, P.A.; Kolber, M. The Reliability, Minimal Detectable Change and Concurrent Validity of a Gravity-Based Bubble Inclinometer and Iphone Application for Measuring Standing Lumbar Lordosis. *Physiother. Theory Pract.* **2013**, *30*, 62–67. [CrossRef]
31. Conceição, F.; Lewis, M.; Lopes, H.; Fonseca, E.M.M. An Evaluation of the Accuracy and Precision of Jump Height Measurements Using Different Technologies and Analytical Methods. *Appl. Sci.* **2022**, *12*, 511. [CrossRef]
32. Kibele, A. Possibilities and Limitations in the Biomechanical Analysis of Countermovement Jumps: A Methodological Study. *J. Appl. Biomech.* **1998**, *14*, 105–117. [CrossRef]
33. Markovic, G. Does Plyometric Training Improve Vertical Jump Height? A Meta-Analytical Review. *Br. J. Sports Med.* **2007**, *41*, 349–355. [CrossRef]
34. Nuzzo, J.L.; Anning, J.H.; Scharfenberg, J.M. The Reliability of Three Devices Used for Measuring Vertical Jump Height. *J. Strength Cond. Res.* **2011**, *25*, 2580–2590. [CrossRef]
35. Chavda, S.; Lake, J.P.; Comfort, P.; Bishop, C.; Joffe, S.A.; Turner, A.N. Relationship Between Kinetic and Kinematic Measures of the Countermovement Jump and National Weightlifting Performance. *J. Sci. Sport Exerc.* **2023**, 1–13. [CrossRef]
36. Janicijevic, D.; Sarabon, N.; Pérez-Castilla, A.; Smajla, D.; Fernández-Revelles, A.; García-Ramos, A. Single-Leg Mechanical Performance and Inter-Leg Asymmetries during Bilateral Countermovement Jumps: A Comparison of Different Calculation Methods. *Gait Posture* **2022**, *96*, 47–52. [CrossRef] [PubMed]
37. Cao, J.; Xun, S.; Zhang, R.; Zhang, Z. Effects of Unilateral, Bilateral and Combined Plyometric Jump Training on Asymmetry of Muscular Strength and Power, and Change-of-Direction in Youth Male Basketball Players. *J. Sports Sci. Med.* **2024**, *23*, 754–766. [CrossRef]
38. Cortis, C.; Tessitore, A.; Lupo, C.; Pesce, C.; Fossile, E.; Figura, F.; Capranica, L. Inter-Limb Coordination, Strength, Jump, and Sprint Performances Following a Youth Men's Basketball Game. *J. Strength Cond. Res.* **2011**, *25*, 135–142. [CrossRef] [PubMed]
39. Ding, L.; Lyu, M.; Chen, Z.; Wu, J.; Wang, Y.; Bishop, C.; Li, Y. Associations Between Inter-Limb Asymmetry in Lower Limb Strength and Jump Performance in 14–15-Year-Old Basketball Players. *Symmetry* **2024**, *16*, 1421. [CrossRef]

applied sciences

MDPI

Article

Intra-Individual Variation in the Jump Smash for Elite Malaysian Male Badminton Players

Yuvaraj Ramasamy [1,*], Yeap Ming Wei [2], Harley Towler [3] and Mark King [3,*]

1 Sports Biomechanics Department, National Sports Institute of Malaysia, Kuala Lumpur 57000, Malaysia
2 Division of Research and Innovation, National Sports Institute of Malaysia, Kuala Lumpur 57000, Malaysia; yeapmingwei@isn.gov.my
3 School of Sport, Exercise and Health Sciences, Loughborough University, Loughborough LE11 3TU, UK; h.towler@lboro.ac.uk
* Correspondence: yuvarajramasamy78@gmail.com (Y.R.); m.a.king@lboro.ac.uk (M.K.)

Abstract: The badminton jump smash is vital for winning rallies, but players can have unintended variations in performance. This study examines the causes of intra-individual smash speed variations among 19 Malaysian male elite players. Twenty-five jump smashes were recorded using a 3D motion capture system, with the three fastest (95.0 ± 3.6 m·s^{-1}) and three slowest (85.8 ± 5.8 m·s^{-1}) smashes per participant categorised into different groups. Six racket/shuttlecock variables and six upper body technique variables during the swing phase and at the point of contact were analysed. Five out of the six racket/shuttlecock variables, including post-impact shuttlecock speed, racket head speed at contact, impact location, distance from the racket's geometric centre, and swing time, significantly differed between groups. Only the shoulder elevation angle differed significantly from 75% to 95% of the forward swing ($p = 0.023$), with less elevation during faster smashes. Consistency in producing high shuttlecock speeds during the jump smash for elite players is primarily achieved through variables linked to timing. Players/coaches should focus on being able to initially produce high racket head speeds, at which point ensuring a consistently efficient impact between the stringbed and shuttlecock should be the priority.

Keywords: SPM analysis; variation in intra-individual; overhead technique; performance racket; jump smash; kinematic time series

1. Introduction

The forehand smash was reported as the last shot played in 29% and 22% of the men's and women's singles rallies at the 2008 Beijing Olympic Games [1]. It is widely regarded as the most effective and aggressive stroke in badminton. The success of this stroke lies in creating a fast downward trajectory of the shuttlecock towards the desired location on the opposite court and putting the opponent in a defensive situation [2]. Unsurprisingly, the post-impact shuttlecock speed is often used to define the quality of the shot. As a result, substantial attention has been paid by researchers to understand the development of speed in badminton smash [3–5], as well as in other similar projectile-hitting sporting motions, such as the tennis serve [6–8] and volleyball jump spike [9–11]. At the elite level, the badminton smash shuttlecock speed may vary due to body positioning, technique, and the timing (impact location) of the smash, among other factors [12]. However, to the authors' knowledge, there is no study to date that has conducted any intra-individual analysis investigating why a given player's shuttlecock speed may vary between smash strokes.

A higher velocity in the smash is associated with specific characteristics at the moment of racket–shuttlecock contact, including a more internally rotated shoulder [3,4], as well as reduced elevation of the shoulder and less extension of the elbow [4]. Additionally, a faster smash is linked to greater angular velocity of the internal rotation of the shoulder, a factor contributing significantly (66%) to the development of racket head speed [13]. Other key elements contributing to the acceleration phase of the smash include forearm pronation [14,15] and wrist flexion [14]. Notably, the rotational motion of the trunk during both the backswing and forward acceleration phases also plays a substantial role in achieving a faster smash speed [5,16]. Furthermore, a consistent impact location of the shuttlecock on the racket face is critical to minimise reductions in smash speed due to the impact location [12].

Previous research has considered differences in technique for overhead sports such as badminton, tennis, and volleyball. In the badminton smash, significantly greater shoulder internal rotation, horizontal abduction, and wrist flexion torque were found for skilled players compared to unskilled players [17]. Furthermore, participants who produce faster smash speeds have been shown to have greater shoulder internal rotation upon impact and a greater pelvis–thorax separation angle in the transverse plane at the end of retraction compared to participants with slower smashes [3]. In the tennis serve, shoulder internal rotation has been shown to be a major contributor to the differences in horizontal racket head velocity between fast and slow servers [8]. In the volleyball spike, differences in upper-limb angular velocities during successful versus faulty spikes were found for different phases, covering 35% to 45% and 76% to 100% of the motion [18].

The parameters that contribute to intra-individual variations in badminton smash speeds are unknown when the goal is to maximise shuttlecock speed. This does not include cases where a reduction in speed may be the goal in favour of a different trajectory, e.g., a steeper smash ('stick smash'). Previously, several key determinants have been identified to differentiate between shuttlecock speeds across a cohort of players; however, these results should not be generalised to an individual [3]. While cohort-based studies provide valuable insights into group trends and biomechanical factors associated with high-performance smashes, they often overlook intra-individual differences. These differences can arise due to unique player-specific techniques, variations in motor control, and adaptations developed through training or anatomical constraints. For instance, individual players may employ distinct smash techniques, including variations in arm kinematics, shoulder elevation, and trunk rotation [5]. They may also display unique sequencing and timing patterns within the kinetic chain, driven by their skill levels, anthropometric characteristics, or personal preferences [2]. As such, the objective of this study was to investigate causes of reduced intra-individual shuttlecock speed during the badminton smash among elite Malaysian players. Understanding whether technique and/or racket/shuttlecock variables cause reductions in shuttlecock speed at an individual level will allow coaches and practitioners to better advise players on how to consistently execute their fastest smash.

2. Materials and Methods

2.1. Participants

Nineteen elite male right-handed Malaysian badminton players were recruited for this study (age: 20.2 ± 0.7 years; mass: 71.7 ± 4.5 kg; height: 1.77 ± 0.06 m; reported as mean ± SD). All players were actively competing in international-level tournaments and representing Malaysia, including BWF-sanctioned events. These players typically rank within the top tiers nationally and have demonstrated consistent performance in competitive settings.

All players were preparing for competitions and injury-free at the time of data collection. Ethical approval was obtained from the National Sports Institute of Malaysia (ISN)

Research Ethics Committee, and written informed consent was obtained from participants prior to their participation. Players' own rackets were used for all testing to reflect real match conditions, with a manufacturer-reported unstrung racket mass of 84.0 ± 1.8 g and self-reported string tensions of 29.8 ± 0.7 lb.

2.2. Data Collection

All data collection was conducted on an instrumented standard full-size badminton court at the Sports Biomechanics Laboratory at the ISN under controlled conditions (Figure 1a). The kinematics of each jump smash were recorded using 25 Oqus 7+ series infrared cameras (Qualisys AB 411 05, Goteborg, Sweden) at 700 Hz (Figure 1a).

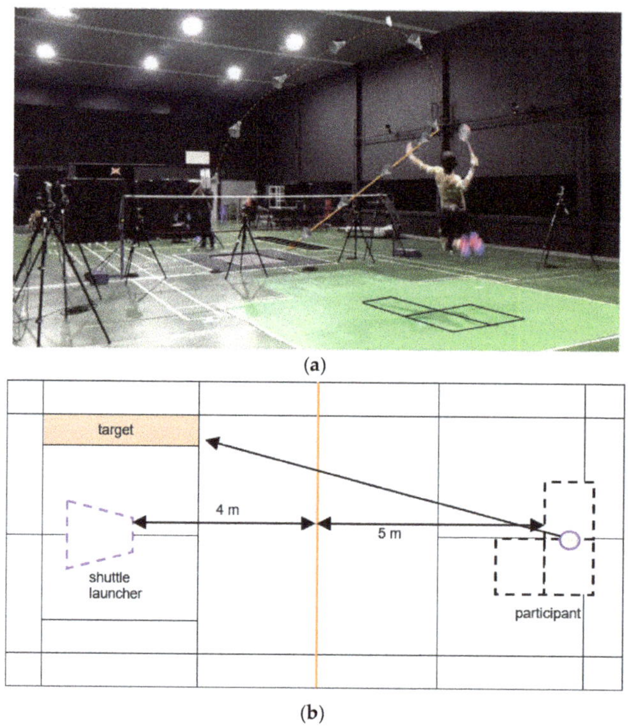

Figure 1. (**a**) Data collection setup, with three force plates and the target area highlighted in black (shuttle path shown from shuttle machine to striking location through to target area); (**b**) setup from above with 3.88 m × 0.4 m target area shown.

A total of sixty-two, 12.5 mm reflective markers were attached to key bony anatomical landmarks (Figure 2a) using Qualisys double-sided and hypafix non-woven adhesive tape to create a 15-segment representation, including the head, thorax, pelvis, 2× upper arms, 2× forearms, 2× hands, 2× thighs, 2× shanks, and 2× feet [4] in Visual 3D v6.00.18 (C-Motion Inc., Germantown, MD, USA). The definition of each segment was constructed from a local reference frame using a minimum of three markers on the segment based on three axes (x: pointing to participants' right; y: pointing forward; z: pointing from distal to proximal structures), as reported in a previous study [3]. Each racket was equipped with one 12.5 mm marker at the bottom of the handle and eight pieces of 10 mm^2 retro-reflective tape on the racket head and shaft (Figure 2b). New Yonex Aerosense 30 shuttlecocks were used for all data collection, with a piece of circle-reflective tape (diameter 19 mm) attached to the tip of the shuttlecock (Figure 2c). Misshaped or broken shuttlecocks were discarded.

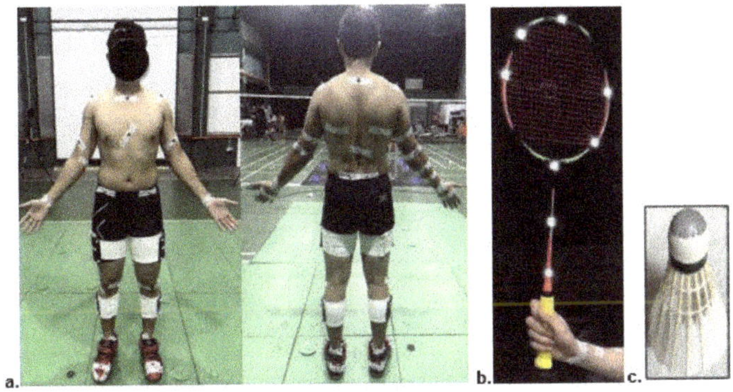

Figure 2. Marker placement on (**a**) the participant, (**b**) the racket, and (**c**) the shuttle tip.

All participants were given approximately 20 min for warm-up and 15–30 min for court familiarisation before data collection. During data collection, high serves/lifts were replicated using a Knight Trainer shuttle launcher (Black Knight, Canada) for the participants to perform their maximum jump smashes, taking off and landing on a fixed area with both feet (Figure 1b). Twenty-five maximum jump smashes aimed at a designated area on the opposite side of the court were performed by each participant. Trials where the shuttlecock did not land in the designated area were excluded.

2.3. Data Reduction

The three fastest and three slowest jump smashes (identified by the post-impact shuttlecock speed) for each participant were categorised into FAST and SLOW conditions, respectively, for further analysis. Position data were labelled using Qualisys Track manager (QTM2021.2; build 4260, Qualisys, Goteburg, Sweden), and gaps within the data were filled using a combination of linear and polynomial interpolation. All body marker position data were filtered within Visual 3D using a low-pass Butterworth filter with a cut-off frequency of 60 Hz, which was determined through a residual analysis [19] using MATLAB R2022a (The MathWorks Inc., Natick, MA, USA).

A 15-segment model in Visual 3D was used to calculate kinematic parameters during the swing phase for each trial according to two key instances: (a) the start, defined as the instance when the racket head speed reached 5 m/s during the backswing phase and did not subsequently drop below this threshold, and (b) shuttle–racket contact, defined by the frame where the anterior–posterior velocity of the shuttlecock changes from negative to positive [20]. Joint angles were calculated as Cardan angles (Table 1), where an xyz rotation sequence was used [21] for all angles except for the shoulder, where a zyz rotation sequence was used [22] following the ISB recommendation, and the thorax, where a yxz rotation sequence was used [23].

Table 1. Cardan joint angle definitions.

Joint	Rotation	Anatomical Position	Positive Direction
thorax	axial rotation (x-factor)	0°	anti-clockwise (viewed from above)
shoulder	plane of elevation	90° [a]	horizontal flexion
	elevation	90° [a]	elevation
	internal/external rotation	180° [a]	internal rotation
elbow	flexion/extension	180°	extension

Note: [a] Shoulder angles are given for the upper arm elevated to the horizontal (abducted by 90° from the anatomical position) and internally rotated such that the forearm is horizontal when the elbow is flexed to 90° (similar to [24], with 90° and 180° added to the plane of elevation and internal rotation angles, respectively, due to the specific ranges of motion in badminton).

Post-impact shuttlecock speed, racket head speed, impact timing (swing time from the start of the forward swing to racket–shuttlecock contact), and impact location were calculated using a curve-fitting methodology [25], which was adapted for badminton [12] to include a 1 ms contact period [26]. Additionally, the racket face angle relative to the vertical upon contact and the shuttlecock angle post-impact relative to the horizontal were calculated [26–28].

2.4. Data Analysis

2.4.1. Technique Variables

The technique variables include the following kinematic parameters: namely, the shoulder plane of elevation, elevation and rotation, elbow flexion/extension angle and angular velocity, and trunk angle data during the swing phase. All angle and angular velocity curves were time-normalised to the start of the forward swing period to the point of racket–shuttle contact. The reliability of each variable for both the FAST and SLOW conditions was analysed using an integrated pointwise index (intraclass correlation, ICC) test.

2.4.2. Racket/Shuttlecock Variables

The post-impact shuttlecock speed, racket head speed, swing time, and impact location (longitudinal and mediolateral) relative to the racket head centre were separated into FAST and SLOW conditions based on the post-impact shuttlecock speed.

2.4.3. Statistical Analysis

Integrated pointwise indices using the ICC (type: consistency) were calculated using variance components of an ANOVA based on a two-way mixed model [29] with pointwise confidence intervals set to 95% supplemented at each time point to analyse the test–retest reliability for the curve data. The classification of ICCs according to another study [30] was adapted for interpretation: poor < 0.4; $0.4 \leq$ fair < 0.6; $0.6 \leq$ good ≤ 0.74; and $0.74 <$ excellent. Data were tested for normality using the Shapiro–Wilk test, where the assumption of normality was met if $p > 0.05$. If a high similarity within each participant's conditions (FAST vs. SLOW) was found (indicated by a high ICC value), the three trials for each condition were averaged (mean) into two single time histories to provide a representative time history of the technique used during each condition. A within-measures ANOVA and statistical parametric mapping (SPM) were performed to assess any significant differences in the technique variables between the conditions over the normalised swing phase. A significance threshold of alpha < 0.05 was set for ANOVA measurements, whilst a Bonferroni correction was used to adjust for post hoc analysis. Paired T-tests were performed in order to detect differences between the racket/shuttlecock variables for each of the two conditions (FAST vs. SLOW). A within-measures ANOVA using 1D statistical parametric mapping (1D-SPM) [31] was performed in MATLAB R2022a (The MathWorks Inc., Natick, MA, USA), and an ICC test and a paired t-test were performed in IBM SPSS Statistics 26 (IBM, Armonk, NY, USA).

3. Results

Overall, the ICC test revealed excellent similarities in all the participants' technique variables in both the FAST and SLOW conditions, identified by a high average ICC value of 0.983 ± 0.065 (mean $\pm$ SD). Hence, the three fastest and three slowest trials for each participant were combined into two single time histories (FAST and SLOW) for each technique variable studied to represent the technique used when performing fast and slow smashes, respectively.

3.1. Technique Variables

Of the five technical variables analysed, namely the shoulder elevation angle (Figure 3), shoulder plane of elevation (Figure 4), shoulder internal/external rotation (Figure 5), elbow flexion/extension (Figure 6), and trunk x-factor angle (Figure 7), no significant differences were observed between the FAST and SLOW conditions during the swing phase, except for the shoulder elevation angle (Figure 3). A significant difference in the shoulder elevation angle was identified from 75% to 95% (Figure 3) of the forward swing between the FAST and SLOW trials ($p = 0.023$), with players being less elevated at the shoulder joint during their FAST trials compared to their SLOW trials.

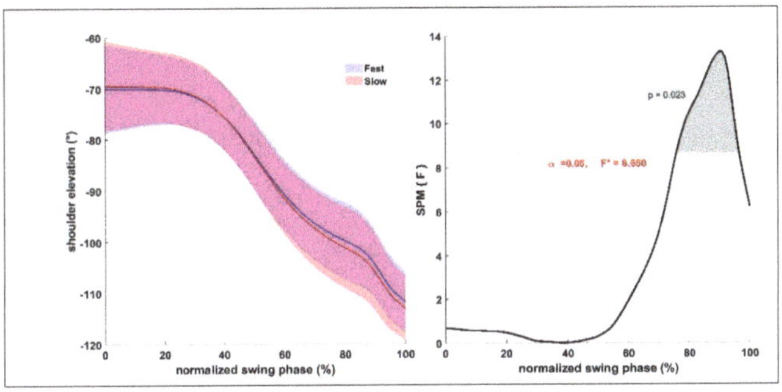

Figure 3. Shoulder elevation angle mean ± standard deviation (SD) for FAST and SLOW conditions on the left and ANOVA SPM (t) on the right.

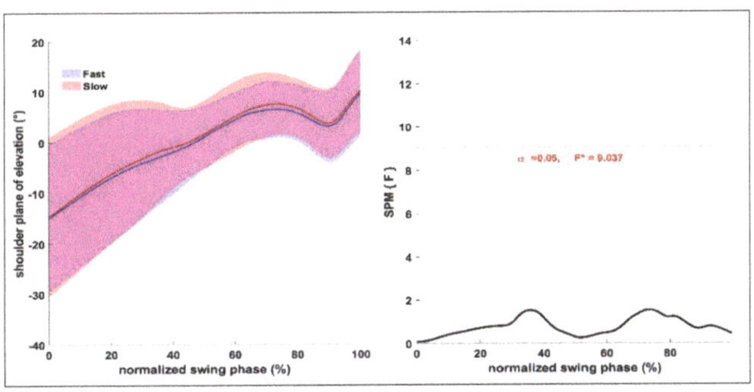

Figure 4. Shoulder plane of elevation angle mean ± standard deviation (SD) for FAST and SLOW conditions on the left and ANOVA SPM (t) on the right.

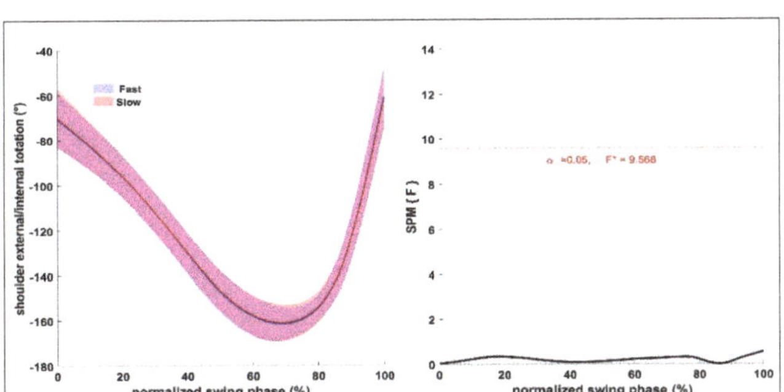

Figure 5. Shoulder internal and external rotation angle mean ± standard deviation (SD) for FAST and SLOW conditions on the left and ANOVA SPM (t) on the right.

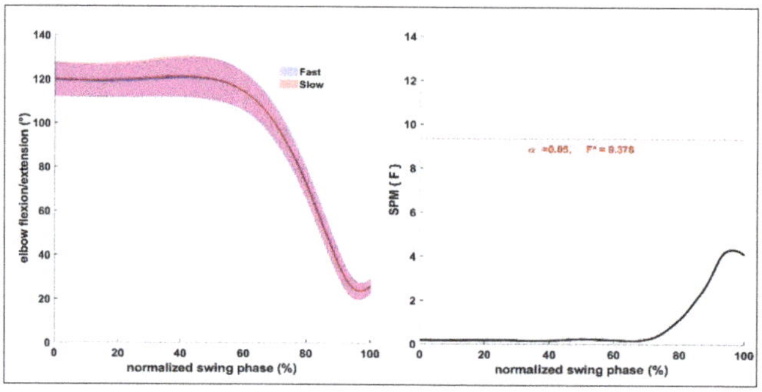

Figure 6. Elbow flexion extension angle mean ± standard deviation (SD) for FAST and SLOW conditions on the left and ANOVA SPM (t) on the right.

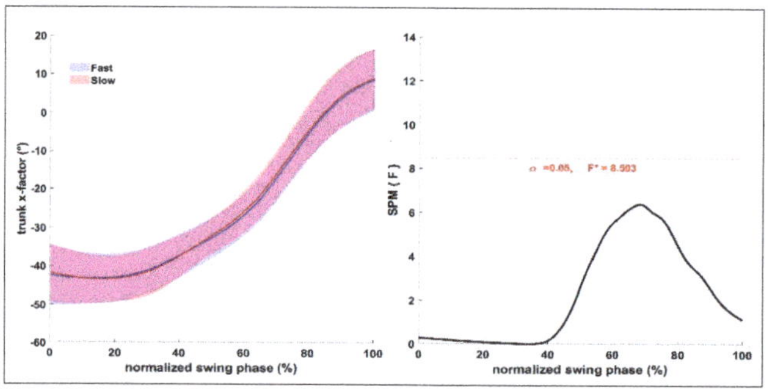

Figure 7. Trunk x- factor angle mean ± standard deviation (SD) for FAST and SLOW conditions on the left and ANOVA SPM (t) on the right.

3.2. Racket/Shuttlecock Variables

The overall range of badminton smash speeds was 75.8–100.6 m·s^{-1}, with the largest variation for an individual being 13.4 m·s^{-1} (13% difference) and the smallest difference for an individual being 3.3 m·s^{-1} (3% difference). Across the 19 participants, significant differences were observed between the FAST and SLOW trials in terms of post-impact shuttlecock speed (95.0 ± 3.6 m·s^{-1} vs. 85.8 ± 5.8 m·s^{-1}, $p < 0.001$), racket head speed at contact (66.3 ± 3.4 m·s^{-1} vs. 64.3 ± 3.7 m·s^{-1}, $p < 0.0005$), and longitudinal impact location (8.5 mm vs. −2.6 mm, $p = 0.027$) (Table 2). Interestingly, the paired T-test also revealed that there were significant differences in the swing duration and impact location between the FAST and SLOW trials. The swing duration during the FAST condition was, on average, 0.0036 s (3.6 ms) shorter than that in the SLOW trials ($p = 0.0012$). Meanwhile, the players were observed to demonstrate 'worse' impact locations in their SLOW trials, demonstrated by the greater absolute distance of the impact location from the geometric centre on average (13.41 mm; $p = 0.001$) compared to in their FAST trials.

Table 2. Paired-samples T-test on the six racket/shuttlecock variables.

Variable	Unit	Mean (SD)		*p*-Value
		FAST	SLOW	
post-impact shuttlecock speed	m·s^{-1}	95.03 ± 3.58	85.79 ± 5.83	<0.0005 *
racket head speed upon contact	m·s^{-1}	66.25 ± 3.43	64.32 ± 3.72	<0.0005 *
mediolateral impact location	mm	−15.31 ± 8.12	−19.57 ± 18.56	0.209

Table 2. *Cont.*

Variable	Unit	Mean (SD)		*p*-Value
		FAST	SLOW	
longitudinal impact location	mm	8.46 ± 15.65	−2.58 ± 27.74	0.027 *
absolute distance of impact location from geometric centre of the racket face	mm	29.48 ± 10.09	42.89 ± 13.76	0.001 *
swing time	ms	162.7 ± 10.5	166.3 ± 12.4	0.001 *

Notes: * denotes statistically significant difference between groups; all results are presented as means (SD).

4. Discussion

The intra-individual variation in shuttlecock smash speed across the 25 trials for each participant varied from 3.3 to 13.4 m·s^{-1}, with significant differences between the FAST and SLOW groups for both post-impact shuttlecock speed and racket head speed. A maximum difference of 3.3 m·s^{-1} (3%) across 25 trials could be considered a small difference and perhaps what might be expected for elite players, but 13.4 m·s^{-1} (13%) is over three times greater than this value and for an elite player is perhaps surprising. This drop of 13% would bring an 'elite' player in line with 'regional'-level players based on data from [3], where elite players achieved similar shuttlecock speeds of 94.6 m·s^{-1} whilst regional players achieved 84.6 m·s^{-1}.

A comparison of technique and racket/shuttlecock variables revealed several important differences between the FAST and SLOW conditions. Unsurprisingly, significant differences existed between the FAST and SLOW conditions for post-impact shuttlecock speed and racket head speed, respectively, given the high correlation between the two variables (r = 0.90; [3]).

Overall the participants performed smashes with greater longitudinal impact locations (11 mm closer to the racket tip) in the FAST group compared to the SLOW group. It is not obvious why this difference in impact location occurred across a group of elite players, but it could indicate that coordinating the timing of the racket shuttle impact was challenging even for elite players. Having an impact location closer to the tip of the racket does result in a faster racket speed at the point of impact on the racket, and therefore, it makes sense that the fastest shuttlecock speeds came from impacts slightly closer to the tip of the racket. There becomes a point at which an increase in longitudinal impact location would cause a reduction in shuttlecock speed due to the impact efficiency between the stringbed and shuttlecock. The mean longitudinal and mediolateral impact locations during the FAST condition were similar to the 'optimal' impact location reported by another study [12], namely 8.5 vs. 11.6 mm and −15.3 vs. −5.7 mm, respectively.

The 'impact quality' may also be related to the significant difference in swing duration, where players performed, on average, 3.6 ms shorter swing durations during the FAST condition. The small amount of extra time taken during the SLOW condition swings perhaps allowed the shuttlecock more time to drop and thus produce impact locations lower on the racket head and further away from the optimal position [12].

The technique of the participants did not vary substantially between the FAST and SLOW conditions, with only one upper-body joint angle differing between the FAST and SLOW conditions, namely the shoulder elevation angle between 75 and 95% of the swing phase. The players held their arm in a less elevated position during the FAST trials compared to the SLOW trials, and this is supported by [4], where the arm is placed in a more favourable position to generate greater angular velocities of shoulder internal rotation. A lower shoulder elevation angle reduces the distance between the racket and the body, effectively shortening the moment arm of the racket–hand system. This potentially reduces the rotational inertia of the arm–racket complex, allowing for faster angular acceleration during the swing phase. The closer proximity of the arm to the body also potentially reduces the torque required to achieve higher racket head speeds, along with the work done against gravity. Given the ability level of the participants, it was unlikely that their technique would differ between the two conditions. Whilst the technique (joint kinematics) used at the various joints remained similar between the conditions, and, therefore, so did the overall ranges of motion, they were completed over shorter periods of time during the FAST condition, implying greater angular velocities and thus greater racket head speeds, transferring more momentum to the shuttlecock (assuming an equivalent impact efficiency).

Previous studies have found shoulder internal rotation upon contact to be an important technique variable, with a more internally rotated shoulder being linked to greater shuttlecock speeds [3,4]. Similarly, players who achieve a more counter-rotated trunk position at the end of the backswing phase have been shown to achieve greater shuttlecock speeds [3–5]. Neither of these technique variables showed any differences between the FAST and SLOW conditions when assessing intra-individual variation throughout the entire swing phase. These results suggest that players changing their technique is unlikely to be the cause of variations in shuttlecock speed, and instead, successfully executing and timing their technique to ensure an efficient impact is likely to explain this variation in shuttlecock speed.

Future work should investigate technique differences with non-elite populations or different age groups, as well as carrying out longitudinal studies, to see what technique changes are made as players progress through various stages and performance groups, thereby highlighting which technique factors are important for consistently developing greater racket head speed and shuttlecock smash speed. It is also likely that players of a lower standard have more variability in their technique, allowing for a greater understanding of what technique factors cause a reduction in shuttlecock speed.

In conclusion, for elite players, the differences in shuttlecock speeds between smash and slow strokes are likely to be caused by racket/shuttlecock variables, not specific changes in technique. In particular, more distal longitudinal impact locations, within sensible limits where impact efficiency is not significantly impacted, and completing the swing phase over a shorter period of time have been shown to generate greater racket head speeds.

5. Limitations

This study's design differentiated between FAST and SLOW smash conditions rather than assigning athletes to distinct groups. This choice may not fully encapsulate motor control or technical variability across different performance levels or fatigue states. All the players in this study were right-handed; the findings should be applicable to left-handed players, but future studies should address this. No specific fatigue assessments were conducted, but rest intervals were provided as required, and there was no specific evidence of fatigue in terms of smash speed, as for some players, their fastest smashes occurred towards the end of the 25 smashes. However, some of the variation found between the conditions could be attributable to factors beyond the designated conditions, such as muscular fatigue. The uniform population of elite male Malaysian players is both a strength and limitation of this study; future studies should look to consider female athletes as well as junior and non-elite participants to broaden the applicability of the findings to the whole population.

6. Conclusions and Recommendations

In summary, this study presents preliminary insights into biomechanical differences in badminton smashes under FAST and SLOW conditions, providing an initial step toward understanding movement variability and performance in sport-specific contexts. Despite the limitations, the findings highlight the importance of nuanced analysis methods in biomechanical studies, particularly in complex, high-performance movements such as the badminton smash. More sophisticated design and analysis choices may be necessary to capture the full scope of the motor control mechanisms underlying elite badminton techniques, especially when addressing intra-individual variations and compensatory joint mechanics. Additionally, from a tactical perspective, it may also be relevant to create conditions where players have a particular focus on generating a maximal shuttlecock speed vs. accuracy balance and evaluate what technique factors change in the likely scenario that players reduce the racket head speed and shuttlecock speed to increase the smash accuracy. A longitudinal study tracking the progression from junior to senior levels could provide valuable insights into the biomechanical changes that occur over time and their impacts on performance. Such an approach would help deepen our understanding of how technique evolves with experience and its influence on smash performance.

Author Contributions: Draft preparation, Y.R.; writing—review and editing, Y.R., Y.M.W., H.T. and M.K.; supervision, Y.R., H.T. and M.K. All authors have read and agreed to the published version of the manuscript.

Funding: This research received internal funding from the National Sports Institute of Malaysia grant number 001/2022-001/2022.

Institutional Review Board Statement: The study was conducted according to the guidelines of the Declaration of Helsinki and approved by the Ethics Committee of the National Sports Institute of Malaysia.

Informed Consent Statement: Informed consent was obtained from all subjects involved in the study.

Data Availability Statement: The data presented in this study are available upon request from the corresponding author. The data are not publicly available due to privacy.

Acknowledgments: The authors would like to express special thanks to the Badminton Association of Malaysia for their support throughout this project.

Conflicts of Interest: The authors declare no conflicts of interest.

References

1. Abian-vicen, J.; Castanedo, A.; Abian, P.; Sampedro, J.; Abian-vicen, J.; Castanedo, A.; Abian, P.; Sampedro, J. Temporal and notational comparison of badminton matches between men' s singles and women' s singles. *Int. J. Perform. Anal. Sport* **2013**, *13*, 310–320. [CrossRef]
2. Phomsoupha, M.; Laffaye, G. The Science of Badminton: Game Characteristics, Anthropometry, Physiology, Visual Fitness and Biomechanics. *Sports Med.* **2014**, *45*, 473–495. [CrossRef] [PubMed]
3. King, M.; Towler, H.; Dillon, R.; McErlain-Naylor, S. A correlational analysis of shuttlecock speed kinematic determinants in the badminton jump smash. *Appl. Sci.* **2020**, *10*, 1248. [CrossRef]
4. Ramasamy, Y.; Sundar, V.; Towler, H.; Usman, J.; King, M.A. Kinetic and kinematic determinants of shuttlecock speed in the forehand jump smash performed by elite male Malaysian badminton players. *Sports Biomech.* **2024**, *23*, 582–597. [CrossRef]
5. Zhang, Z.; Li, S.; Wan, B.; Visentin, P.; Jiang, Q.; Dyck, M.; Li, H.; Shan, G. The Influence of X-Factor (Trunk Rotation) and Experience on the Quality of the Badminton Forehand Smash. *J. Hum. Kinet.* **2016**, *53*, 9–22. [CrossRef]
6. Gordon, B.J. Contributions of joint rotations to racquet speed in the tennis serve. *J. Sports Sci.* **2006**, *24*, 31–49. [CrossRef]
7. Koike, S.; Hashiguchi, T. Dynamic contribution analysis of badminton-smash-motion with consideration of racket shaft deformation (A model consisted of racket-side upper limb and a racket). *Procedia Eng.* **2014**, *72*, 496–501. [CrossRef]
8. Tanabe, S.; Ito, A. A three-dimensional analysis of the contributions of upper limb joint movements to horizontal racket head velocity at ball impact during tennis serving. *Sports Biomech.* **2007**, *6*, 37–41. [CrossRef]
9. Fuchs, P.X.; Menzel, H.J.K.; Guidotti, F.; Bell, J.; von Duvillard, S.P.; Wagner, H. Spike jump biomechanics in male versus female elite volleyball players. *J. Sports Sci.* **2019**, *37*, 2411–2419. [CrossRef]
10. Seminati, E.; Marzari, A.; Vacondio, O.; Minetti, A.E. Shoulder 3D range of motion and humerus rotation in two volleyball spike techniques: Injury prevention and performance. *Sports Biomech.* **2015**, *14*, 216–231. [CrossRef]

11. Serrien, B.; Ooijen, J.; Goossens, M.; Baeyens, J.-P. A Motion Analysis in the Volleyball Spike—Part 1: Three-dimensional Kinematics and Performance. *Int. J. Hum. Mov. Sports Sci.* **2016**, *4*, 70–82. [CrossRef]
12. McErlain-Naylor, S.A.; Towler, H.; Afzal, I.A.; Felton, P.J.; Hiley, M.J.; King, M.A. Effect of racket-shuttlecock impact location on shot outcome for badminton smashes by elite players. *J. Sports Sci.* **2020**, *38*, 2471–2478. [CrossRef] [PubMed]
13. Liu, X.; Kim, W.; Tan, J. *An Analysis of the Biomechanics of Arm Movement During a Badminton Smash*; Nanyang Technology University: Singapore, 2002; Volume 1.
14. Rusdiana, A.; Ruhayati, Y.; Korea, S.; Korea, S. 3D Kinematic Analysis of Standing and Jumping Smash Technique of Indonesian Badminton National Elite Athletes. *Man India* **2016**, *96*, 2525–2535.
15. Zhang, Z. The Influence of Body Positioning, Trunk Rotation (X-Factor) and Training Effect on Quality of the Badminton Forehand Overhead Smash. Master's Thesis, University of Lethbridge, Lethbridge, AB, Canada, 2012.
16. Teu, K.K.; Kim, W.; Tan, J.; Fuss, F.K. Using dual Euler angles for the analysis of arm movement during the badminton smash. *Sports Eng.* **2005**, *8*, 171–178. [CrossRef]
17. Rusdiana, A.; Bin Abdullah, M.R.; Syahid, A.M.; Haryono, T.; Kurniawan, T. Badminton overhead backhand and forehand smashes: A biomechanical analysis approach. *J. Phys. Educ. Sport* **2021**, *21*, 1722–1727. [CrossRef]
18. Sarvestan, J.; Svoboda, Z.; Alaei, F.; Mulloy, F. Analysis of Whole-Body Coordination Patterning in Successful and Faulty Spikes Using Self-Organising Map-Based Cluster Analysis: A Secondary Analysis. *Sensors* **2021**, *21*, 1345. [CrossRef]
19. Winter, D.A. *Biomechanics and Motor Control of Human Movement*; Wiley: Hoboken, NJ, USA, 2009.
20. Ramasamy, Y.; Sundar, V.; Usman, J.; Towler, H.; King, M.A. Relationships between Racket Arm Joint Moments and Racket Head Speed during the Badminton Jump Smash Performed by Elite Male Malaysian Players. *Appl. Sci.* **2022**, *12*, 880. [CrossRef]
21. Worthington, P.J.; King, M.A.; Ranson, C.A. Relationships between fast bowling technique and ball release speed in cricket. *J. Appl. Biomech.* **2013**, *29*, 78–84. [CrossRef]
22. Wu, G.; Van Der Helm, F.C.T.; Veeger, H.E.J.; Makhsous, M.; Van Roy, P.; Anglin, C.; Nagels, J.; Karduna, A.R.; McQuade, K.; Wang, X.; et al. ISB recommendation on definitions of joint coordinate systems of various joints for the reporting of human joint motion—Part II: Shoulder, elbow, wrist and hand. *J. Biomech.* **2005**, *38*, 981–992. [CrossRef]
23. Smith, A.C.; Roberts, J.R.; Wallace, E.S.; Kong, P.; Forrester, S.E. Comparison of Two- and Three-Dimensional Methods for Analysis of Trunk Kinematic Variables in the Golf Swing. *Appl. Biomech.* **2015**, *32*, 23–31. [CrossRef]
24. Williams, B.K.; Sanders, R.H.; Ryu, J.H.; Graham-Smith, P.; Sinclair, P.J. The kinematic differences between skill levels in the squash forehand drive, volley and drop strokes. *J. Sports Sci.* **2020**, *38*, 1550–1559. [CrossRef] [PubMed]
25. Peploe, C.; McErlain-Naylor, S.A.; Harland, A.R.; King, M.A. The relationships between impact location and post-impact ball speed, bat torsion, and ball direction in cricket batting. *J. Sports Sci.* **2018**, *36*, 1407–1414. [CrossRef] [PubMed]
26. Cohen, C.; Texier, B.D.; Quéré, D.; Clanet, C. The physics of badminton. *New J. Phys.* **2015**, *17*, 063001. [CrossRef]
27. Li, S.; Zhang, Z.; Wan, B.; Wilde, B.; Shan, G. The relevance of body positioning and its training effect on badminton smash. *J. Sports Sci.* **2016**, *35*, 310–316. [CrossRef]
28. Zhu, Q. Expertise of using striking techniques for power stroke in badminton. *Percept. Mot. Skills* **2013**, *117*, 427–441. [CrossRef]
29. Schelin, L.; Pini, A.; Markström, J.L.; Häger, C.K. Test-retest reliability of entire time-series data from hip, knee and ankle kinematics and kinetics during one-leg hops for distance: Analyses using integrated pointwise indices. *J. Biomech.* **2021**, *124*, 110546. [CrossRef]
30. Cicchetti, D.V. Guidelines, Criteria, and Rules of Thumb for Evaluating Normed and Standardized Assessment Instruments in Psychology. *Psychol. Assess.* **1994**, *6*, 284–290. [CrossRef]
31. Pataky, T.C. Generalized n-dimensional biomechanical field analysis using statistical parametric mapping. *J. Biomech.* **2010**, *43*, 1976–1982. [CrossRef]

applied sciences

MDPI

Article

Enhancing Acceleration Capabilities in Professional Women's Football Players: A Comparative Analysis of Game-Based Versus Resisted Sprint Trainings

Adrián Castaño-Zambudio [1,*], Carmen Repullo [2] and Pedro Jiménez-Reyes [1]

1 Sport Sciences Research Centre, Rey Juan Carlos University, 28943 Fuenlabrada, Madrid, Spain
2 Education Faculty, Autonomous University of Madrid, 28049 Madrid, Spain
* Correspondence: adrian.castano@urjc.es

Featured Application: Both integrated game-based and resisted sprint training methods effectively enhance acceleration in professional women's football. Incorporating high-speed actions into training significantly improves sprint performance across distances, offering a practical solution during competitive periods.

Abstract: The recognition of high-speed demands in football has led elite academies to prioritize acceleration capabilities for player selection and promotion, particularly given their fundamental role in the motor skills of professional players and their impact on goal-related opportunities. This study explored the effectiveness of game-based versus resisted sprint training methods in enhancing the acceleration abilities of professional women's football players. Over the entire competitive period, the training load of 26 athletes (24.2 ± 3.7 years) was assessed using GPS devices, and sprint capabilities were evaluated through four 30-m acceleration tests spaced six weeks apart. Linear mixed models (LMMs) analyzed physical load parameters, including distance covered at high speeds, speed events, and maximum speed, with periods and players as fixed and random effects, respectively. Significant sprint performance improvements were observed across all intervals, particularly when high-intensity distance volumes were combined with resisted sprint training. Conversely, high-intensity running without additional stimuli also led to performance gains, albeit to a lesser extent. Both game-based and resisted sprint training methods were effective in enhancing acceleration capabilities, while the absence of specific sprint focus did not significantly alter sprint performance. These findings support the inclusion of tailored sprint training in athletic programs to optimize acceleration in women's football players.

Keywords: sprint performance; team sports; women's football; training load; resisted sprint

1. Introduction

In contemporary women's football, high-intensity conditional demands have increased significantly [1], mirroring trends in their male counterparts [2,3]. Such demands play a crucial role in talent recruitment, especially in elite academies, with acceleration capabilities becoming pivotal criteria for player selection and promotion [4–6].

Acceleration is not only integral to a football player's skill set but also crucial for athletes in team-based sports, particularly in performance-driven scenarios [7–10]. In football, acceleration plays a fundamental role in winning duels, defending, and creating goal-scoring opportunities [7]. Sprinting, as one of the most frequent actions leading to goals, highlights its importance in these decisive moments [7,10]. The ability to accelerate effectively provides players with a competitive edge in game-critical situations, reinforcing the necessity of developing this skill as part of an athlete's overall performance profile. Moreover, these abilities have been found to differentiate athletes based on their sporting levels [11,12]. This recognition has increased interest among both practitioners and researchers, leading to more studies on improving individual acceleration [13–16]. However, some experts suggest that selecting naturally faster players may be more effective than trying to improve players' acceleration capabilities [17].

To address this objective, different approaches have targeted the specific enhancement of these capabilities. Sprint-specific methodologies adopt a decontextualized approach, prioritizing targeted athlete stimulation over game-related contextual factors. This allows for precise performance improvements without the interference of in-game variables [14–16,18]. Among these methodologies, resisted sprint training has gained popularity in team sports for its effectiveness in improving early acceleration under diverse overloading conditions [18]. In contrast, free and assisted sprinting are more suited to phase-specific adaptations, particularly in optimizing force application at high speeds [15]. Collectively, these sprint-specific interventions—whether resisted, free, or assisted—have been shown to improve sprint performance across various sports and competition levels [19–21]. Yet, the significance of exerting effort in relevant contexts is likely to contribute to the growing adoption of game-based methods [22]. Among these, small-sided games (SSGs) have gained prominence for their emphasis on technical–tactical skills and frequent short, high-intensity bursts. Nevertheless, although they are effective at improving cardiovascular fitness and mimicking official match acceleration–deceleration patterns, some studies indicate they might not optimally increase high-speed running or mirror the most intense competition periods [23,24]. Conversely, recent research suggests that adjustments in factors such as the number of players, individual player space, and space organization can potentially bridge the disparities in demands associated with high-speed activities [25–28].

Optimizing these abilities remains a priority in sports research and practice. This is not only due to their significant impact on player performance in sports but also due to the growing connections to the most frequent football injury, the hamstring tear [29–34]. We hypothesize that (i) the acceleration capabilities of female professional football players can be optimized during the competitive period, and (ii) both approaches can improve these capabilities, provided that the stimulus is delivered correctly. This study aimed to assess the effectiveness of integrated and resisted sprint training protocols in improving acceleration ability in professional women's football.

2. Materials and Methods

2.1. Subjects

Twenty-six professional female football players from a single Spanish first division team, with an average age of 24.2 years (±3.7), a height of 1.69 m (±0.07), and a top speed of 27.2 km/h (±2.8), volunteered for this study. These athletes engaged in a regular weekly exercise routine that consisted of five football practices (approximately 8 h in total), five strength/power exercise sessions, and one competitive match per weekend. Notably, none of the players had prior experience with sprint-specific training. Their distribution across positions was as follows: 8 defenders, 8 midfielders, 6 wingers, and 4 forwards. Goalkeepers were excluded from the study due to distinct variations in their preparatory periods compared to outfield players. A priori power analysis was conducted to ensure the study's ability to detect significant changes in sprint performance (0–30 m) over time. A linear mixed model (LMM) was fitted, with testing period (T1–4) as a fixed effect and random intercepts for each player ('player ID'), accounting for within-subject variability. Based on 100 simulations with the current sample size of 26 players and 127 total observations, the power was estimated at 87.0% (95% CI [81.3, 92.4]), confirming that the sample size was sufficient to detect significant differences between periods. Prior to commencing the investigation, written informed consent was obtained from the players, and where needed, from their parents. This study protocol, in accordance with the principles of the Declaration of Helsinki, was granted approval by the research ethics committee at the Universidad Rey Juan Carlos (Ref: 0102202306623).

2.2. Study Design and Procedures

A longitudinal quasi-experimental design was employed to understand the effects of different approaches to stimulating acceleration capabilities at various points during the regular season. Participating players underwent three 6-week training periods during the regular season. These periods were divided into (i) regular football training (integrated) supplemented with resisted sprint-specific training (IRSST); (ii) integrated training without a focus on high-speed actions (INT); and (iii) integrated training with a focus on high-speed actions (INTHS). Regular football practice involved 1 active recovery session (or complementary training for players who participated less than sixty minutes in the previous match); 1 passive recovery session (no training at all); and 3 optimization training sessions characterized by (i) short, high-intensity events and neuromuscular actions through relatively dense (<100 m^2 per player) football drills with a lower number of players; (ii) tactically oriented football drills characterized by larger individual areas (>200 m^2 per player) and a greater number of players involved; and (iii) low-volume, high-intensity training sessions where high-speed events are interspersed with long rest periods. Lastly, pre-match training consisted of high-intensity, short-distance, high work-to-rest drills. The schematic representation of the experimental design is illustrated in Figure 1.

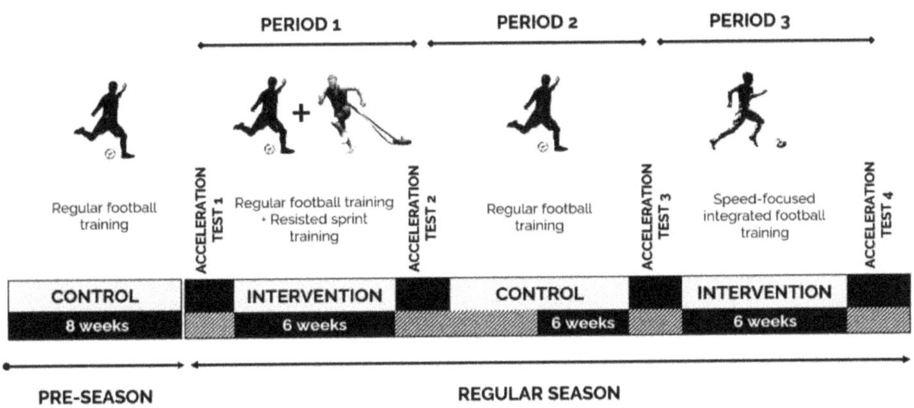

Figure 1. Schematic representation of the experimental design.

2.3. Sprint Training

Two training protocols were designed to complement game-based approaches and were present throughout the 3 intervention periods. The structure and characteristics of the game-based methodology were kept constant across the different intervention periods. The training for period 1 (P1) consisted of sprint-specific training characterized by a progressive overload in both intensity and volume of training. A detailed description can be found in Table 1. The training for period 2 (P2) consisted of its usual training dynamics, characterized by a game-based methodology. Finally, the training for period 3 (P3) maintained the same structure as that present in P2, complemented with game-based tasks that emphasized high-speed situations through larger relative areas per player (>200 m^2 per player) or the inclusion of specific tasks aimed at recreating transition situations.

Table 1. Resisted training program training.

Session/Week	Training		Volume
1	3 × 20 m 20% BM	2 × 20 m 25% BM	100 m
2	4 × 20 m 30% BM	2 × 20 m 25% BM	120 m
3	5 × 20 m 40% BM	2 × 20 m 50% BM	140 m
4	6 × 20 m 60% BM	2 × 20 m 50% BM	160 m
5	7 × 20 m 70% BM	2 × 20 m 75% BM	180 m
6	8 × 20 m 80% BM	2 × 20 m 75% BM	200 m

Abbreviations: BM—body mass; and m—meters.

2.4. Sprint Performance Testing

The acceleration capabilities of the athletes were assessed using a 30-m maximum acceleration test starting from stationary. These capabilities were calculated based on the mean values of two 30-m sprint tests. Assessments were conducted at four time points throughout the season: T1 (after the pre-season), T2 (post-resisted intervention), T3 (post-control period), and T4 (post-integrated intervention). Data were collected using timing gates (Witty Microgate, Microgate, Bolzano, Italy): Dual-beam timing gates were placed on the track 1 m above ground level at 0, 20, and 30 m from the starting line to monitor training sessions of free sprinting. The starting position was located 0.5 m behind the first timing gate. A visual representation of a sprint testing session is shown in Figure 2. All tests were preceded by a standardized warm-up (6 min running at 8–10 km/h with general conditioning exercises and dynamic stretching) and a specific warm-up including wall drills, progressive speed runs, and resisted and flying runs. Likewise, all tests were performed at the same time of day for each player in order to avoid any negative effects of circadian rhythms. Verbal encouragement was provided by the research staff during all tests.

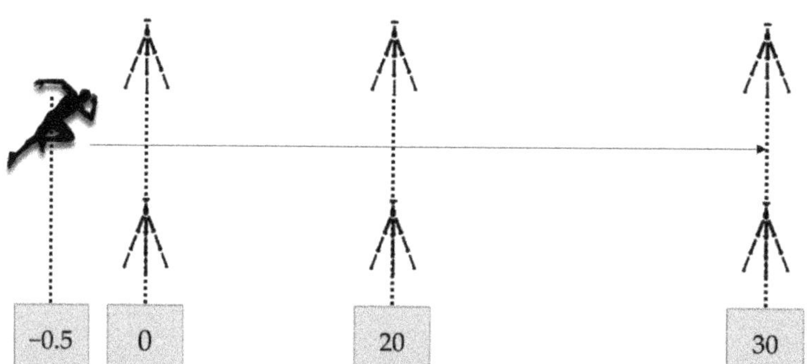

Figure 2. Sprint testing session setup. Starting position is located at −0.5 m, with testing distances labeled at 0, 20 and 30 m.

2.5. External Load Monitoring

Daily external load (EL) records were collected using GPS units at a rate of 18 Hz (GPEXE Pro2, Exelio SRL, Udine, Italy, firmware version 0.13) and analyzed for the 6 weeks leading up to each assessment (with the exception of the pre-season measurement). Sprint-related variables, such as (i) the number of meters covered at high (>21 km/h) and (ii) very high speeds (>24 km/h); (iii) the number of speed events (actions involving > 1 s runs above 21 km/h); and (iv) the quantity of accelerations (>3 m/s^2), were considered both in absolute terms and relative to the practice time during the three periods between measurements: (a) post pre-season (T1) to post-resisted intervention (T2); (b) post-resisted intervention (T2) to post-control period (T3); and (c) post-control period (T3) to post-integrated intervention (T4). Along with these, the maximum capability of (v) acceleration and the (vi) peak speed were included in the analysis. Variables such as the distance covered by walking or jogging were excluded from the analysis due to their limited significance in optimizing these capabilities.

2.6. Statistical Analysis

Statistical analyses were conducted using R-Studio for Windows (version 4.4.1; Posit, PBC, Boston, MA, USA, 2020) with the lme4 package, as well as with jamovi (Version 2.3) [Computer Software]. The Kolmogorov–Smirnov test and Levene's test were employed to assess the normality of residuals and homogeneity of variance, respectively. Both the performance-related and EL data exhibited a normal distribution and homogeneous variance. We employed LMMs, estimated using REML and the nloptwrap optimizer, to understand the relationship between sprint performance across different distances and periods. For predictions related to 0–30 m, 0–20 m, and 20–30 m distances based on the period (represented respectively by the formulas: 0–30 m~1 + period, 0–20 m~1 + period, and 20–30 m~1 + period), we incorporated 'player ID' as a random effect to account for individual player variability. The same was replicated for EL parameters with absolute and relative time distances above 21 and 24 km/h, speed events, and maximum speed for the defined periods in order to estimate the average workload per session during the analyzed period. The fixed factors comprised 6-week periods (P1–3) for EL sprint-related parameters and testing sessions (T1–4) for distinct assessment time points. Random intercepts were assigned to individual players in both analyses. The confidence level was established at 95%, and values of $p < 0.05$ were considered statistically significant. Upon determining the significance of the fixed factor, Bonferroni post hoc pairwise comparisons were employed.

The adoption of this analysis approach is recommended over more traditional designs, such as repeated measures ANOVA, due to its ability to handle incomplete designs and account for correlations from repeated observations across various microcycles. LMMs also effectively manage missing data without requiring imputation or case exclusion, thus retaining all available data, minimizing bias, and enhancing the reliability of results. This comprehensive approach ensures the dataset's integrity and provides a more robust analysis, as supported by recent sports science research [35].

3. Results

3.1. Sprint Performance

For the 0–30 m prediction, the model demonstrated a significant overall explanatory power, with a conditional r^2 of 0.89 and a marginal r^2 of 0.12. The intercept, when the period equals zero, was at 4.86 (95% CI [4.76, 4.96], t(59) = 96.70, $p < 0.001$). Similarly, for the 0–20 m prediction, the model displayed a conditional r^2 of 0.85, a marginal r^2 of 0.08, and an intercept at 3.46 (95% CI [3.39, 3.54], t(59) = 92.03, $p < 0.001$). Finally, the model predicting the 20–30 m range showcased a conditional r^2 of 0.84, a marginal r^2 of 0.22, and an intercept at 1.40 (95% CI [1.37, 1.43], t(59) = 83.00, $p < 0.001$).

Significant main effects were observed for the testing session across three intervals: 0–30 m (beta = −0.07, t(59) = −7.64, $p < 0.001$), 0–20 m (beta = −0.04, t(59) = −5.17, $p < 0.001$), and 20–30 m (beta = −0.03, t(59) = −8.38, $p < 0.001$). These findings suggest that the effect size was consistently negative across all intervals, with standardized beta values indicating the following magnitudes: for 0–30 m, Std. beta = −0.35, 95% CI [−0.44, −0.26]; for 0–20 m, Std. beta = −0.28, 95% CI [−0.39, −0.17]; and for 20–30 m, Std. beta = −0.46, 95% CI [−0.57, −0.35] (see Figure 3).

Specifically, post hoc analysis revealed significant differences between T1 and the other testing sessions ($p < 0.001$) across all analyzed distances (0–30 m, 0–20 m, and 20–30 m). Additionally, T2 and T3 were found to differ significantly from T4. However, no statistical differences were observed between the control period (T2 vs. T3) (see Table 2).

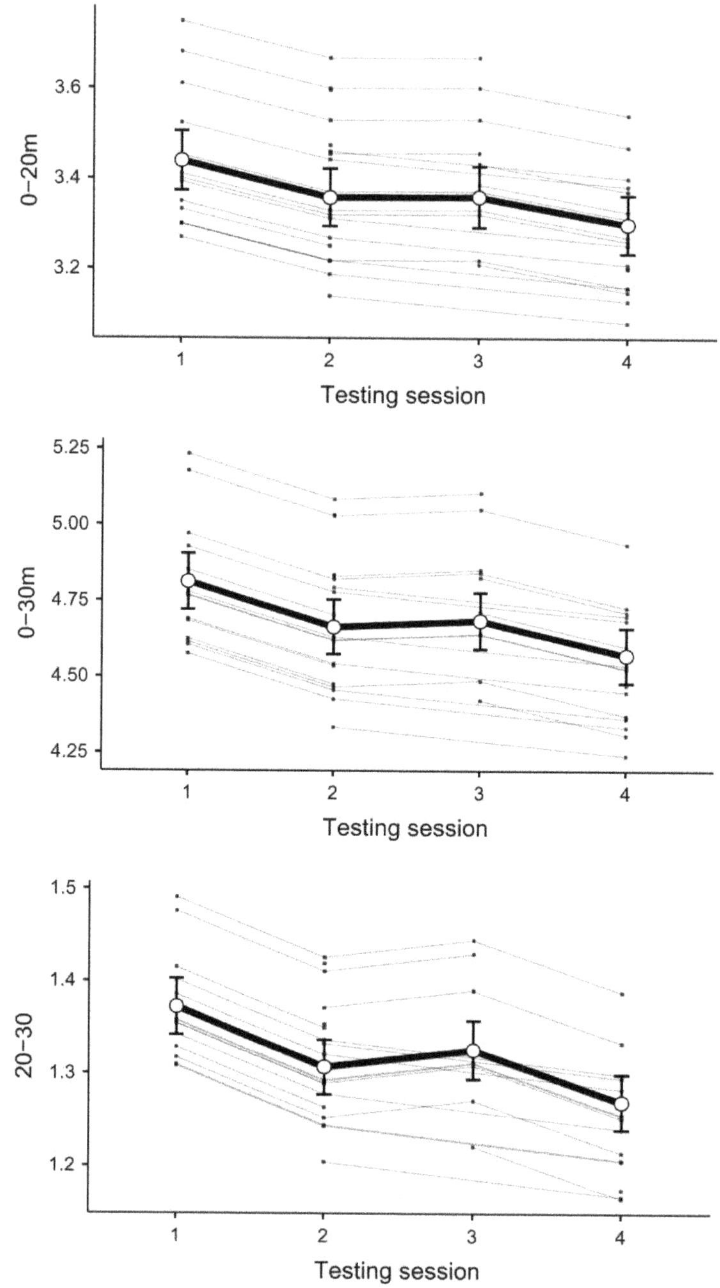

Figure 3. Sprint performance for 0–20 m, 0–30 m, and 20–30 across different testing sessions.

Table 2. Comprehensive post hoc analysis of sprint performance across testing sessions.

	Sprint Performance				
	1st Test	**2nd Test**	**3rd Test**	**4th Test**	**Post Hoc**
0–20	3.44 ± 0.03	3.36 ± 0.04	3.35 ± 0.03	3.3 ± 0.03	T1–T2: $p < 0.001$; Diff: 0.08; t: 4.82 T1–T3: p 0.002; Diff: 0.08; t: 7.67 T1–T4: $p < 0.001$; Diff: 0.14; t: 7.67 T2–T4: p 0.005; Diff: 0.06; t: 3.43 T3–T4: p 0.019; Diff: 0.06; t: 3.02
					T1–T2: $p < 0.001$; Diff: 0.15; t: 7.77 T1–T3: $p < 0.001$; Diff: 0.13; t: 5.33 T1–T4: $p < 0.001$; Diff: 0.24; t: 11.61 T2–T4: $p < 0.001$; Diff: 0.09; t: 4.74 T3–T4: $p < 0.001$; Diff: 0.11; t: 5.00
20–30	1.37 ± 0.01	1.31 ± 0.01	1.32 ± 0.02	1.27 ± 0.02	T1–T2: $p < 0.001$; Diff: 0.07; t: 7.19 T1–T3: $p < 0.001$; Diff: 0.05; t: 4.17 T1–T4: $p < 0.001$; Diff: 0.10; t: 10.63 T2–T4: $p < 0.001$; Diff: 0.04; t: 4.07 T3–T4: $p < 0.001$; Diff: 0.06; t: 5.31

Notes: T1, T2, T3, and T4 represent the first, second, third, and fourth testing sessions, respectively; Diff—absolute difference(s) between tests; and *p*—*p*-value.

3.2. External Load Analysis

LMMs were fitted to predict different metrics using the period variable, consistently treating 'player ID' as a random effect. The model targeting distance/Z5(m) showed a conditional r^2 of 0.13, a marginal r^2 of 4.61×10^{-3}, and an intercept of 38.45 (95% CI [26.02, 50.88], t(1725) = 6.07, $p < 0.001$). Whereas for distance/Z5(m)′, the conditional r^2 was 0.03, the marginal r^2 was 0.03, and the intercept was −0.12 (95% CI [−0.49, 0.25], t(1725) = −0.64, $p = 0.521$). In the model predicting distance/Z6(m) using the formula distance/Z6(m) ~ 1 + period, the model displayed moderate explanatory power (conditional $R^2 = 0.14$) and a marginal r^2 of 4.89×10^{-3}. Its intercept, corresponding to period = 0, was valued at 10.98 (95% CI [2.06, 19.90], t(1725) = 2.41, $p = 0.016$). In contrast, the model for distance/Z6(m)′ revealed a conditional r^2 of 0.03 and a marginal r^2 of 0.02, with an intercept at −0.24 (95% CI [−0.53, 0.05], t(1725) = −1.61, $p = 0.108$). Predicting "max speed" (km/h) led to a model with a conditional r^2 of 0.12, marginal r^2 of 4.22×10^{-3}, and an intercept of 23.10 (95% CI [22.52, 23.68], t(1725) = 78.01, $p < 0.001$). For "speed events", the model's explanatory power was moderate (conditional $R^2 = 0.17$), with a marginal r^2 of 7.31×10^{-3}, and an intercept of 15.02 (95% CI [11.73, 18.30], t(1725) = 8.97, $p < 0.001$). Lastly, the model pertaining to "speed events′" exhibited a conditional r^2 of 0.06, a marginal r^2 of 0.04, and an intercept of 0.03 (95% CI [−0.05, 0.12], t(1725) = 0.81, $p = 0.419$).

The period variable had significant effects on the z5 distance (beta = 5.73, t(1725) = 2.80, $p = 0.005$) and on the z5′ distance (beta = 0.54, t(1725) = 6.61, $p < 0.001$), followed by pronounced effects on the z6 distance (beta = 4.14, t(1725) = 2.90, $p = 0.004$) and on the z6′ distance (beta = 0.34, t(1725) = 5.36, $p < 0.001$). This illustrates that the effect size was consistently positive across these distances, with standardized beta values providing further clarity: for z6, Std. beta = 0.07, 95% CI [0.02, 0.12]; for z6′, Std. beta = 0.13, 95% CI [0.08, 0.18]; for z5, Std. beta = 0.07, 95% CI [0.02, 0.12]; and for z5′, Std. beta = 0.16, 95% CI [0.11, 0.21] (see Figure 4).

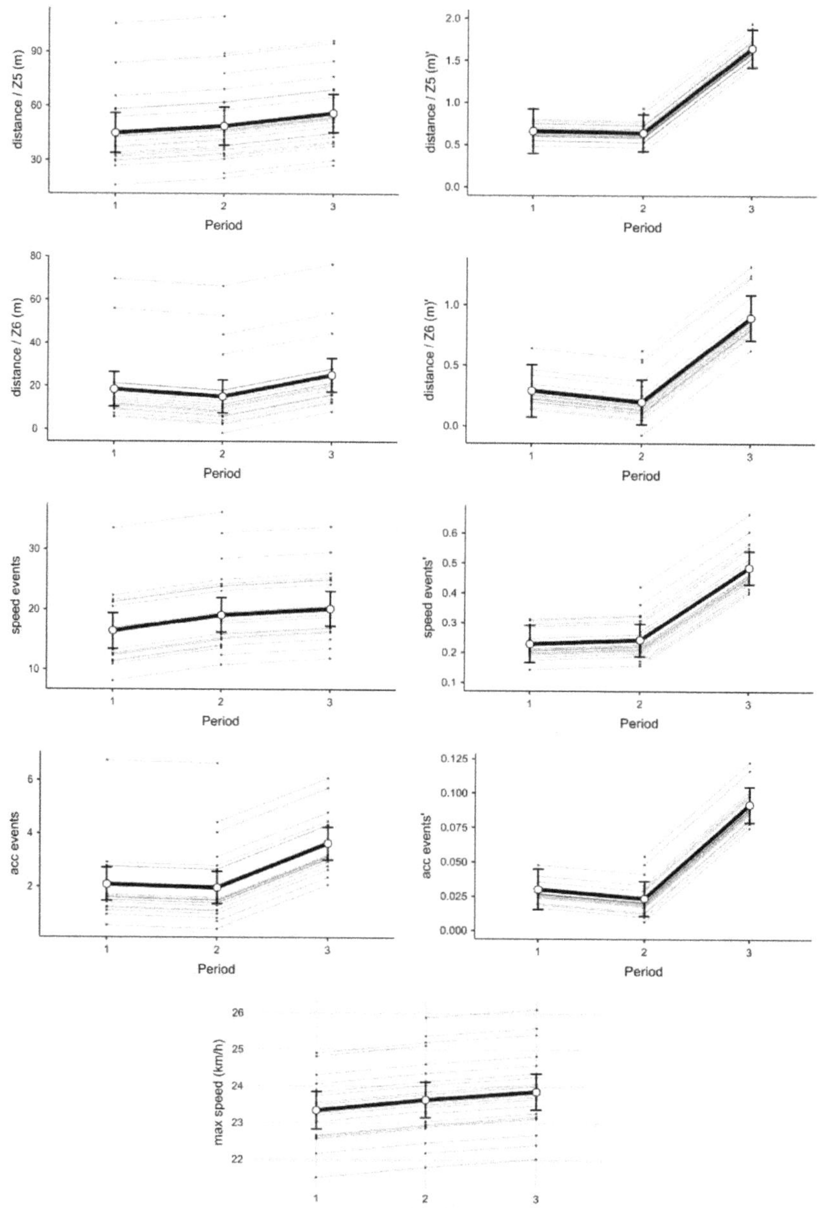

Figure 4. Visual representation of several key high-intensity parameters analyzed over the periods.

For high-speed running distances, post hoc analysis found statistically significant differences in absolute volumes between P1 and P3. Similarly, differences in relative volumes were significant for both the P1–P3 and P2–P3 intervals ($p < 0.001$) (see Table 3).

Table 3. Comprehensive post hoc analysis of absolute and relative workload parameters at high intensity, alongside principal workload indicators by period.

		Period 1	Period 2	Period 3	Post Hoc
Absolute Workload	distance/Z5 (m)	45.0 ± 5.45	48.9 ± 5.14	56.1 ± 5.21	P1–P3: p: 0.024; Diff: −11.1; t: −2.65
	distance/Z6 (m)	18.4 ± 3.95	15.1 ± 3.75	25.1 ± 3.79	P2–P3: $p < 0.001$; Diff: −10; t: −4.19
	speed events	16.4 ± 1.47	19.1 ± 1.41	20.2 ± 1.43	P1–P2: p: 0.021; Diff: −2.66; t: −2.71; P1–P3: $p < 0.001$; Diff: −3.75; t: −3.72
	acc events	2.09 ± 0.31	1.96 ± 0.30	3.64 ± 0.30	P1–P3: $p < 0.001$; Diff: −1.55; t: −7.36; P2–P3: $p < 0.001$; Diff: −1.67; t: −9.66
Relative Workload	distance/Z5 (m)′	0.66 ± 0.13	0.64 ± 0.11	1.65 ± 0.11	P1–P3: $p < 0.001$; Diff: −0.99; t: −5.99 P2–P3: $p < 0.001$; Diff: −1.0; t: −7.0
	distance/Z6 (m)′	0.29 ± 0.11	0.20 ± 0.09	0.90 ± 0.09	P1–P3: $p < 0.001$; Diff: −0.61; t: −4.72; P2–P3: $p < 0.001$; Diff: −0.7; t: −6.31
	speed events′	0.229 ± 0.03	0.243 ± 0.03	0.486 ± 0.03	P1–P3: $p < 0.001$; Diff: −0.26; t: −7.33; P2–P3: $p < 0.001$; Diff: −0.24; t: −8.11
	acc events′	0.03 ± 0.01	0.02 ± 0.01	0.09 ± 0.01	P1–P3: $p < 0.001$; Diff: −0.06; t: −6.95; P2–P3: $p < 0.001$; Diff: −0.06; t: −8.91
Intensity Indicators	max speed (km/h)	23.3 ± 0.25	23.6 ± 0.24	23.9 ± 0.24	P1–P3: p: 0.023; Diff: −0.52; t: −2.66
	max acc (m/s^2)	3.91 ± 0.04	3.94 ± 0.04	4.05 ± 0.04	P1–P3: $p < 0.001$; Diff: −0.15; t: −4.09 P2–P3: $p < 0.001$; Diff: −0.11; t: −3.7
	max dec (m/s^2)	−4.30 ± 0.06	−4.46 ± 0.05	−4.50 ± 0.05	P1–P2: $p < 0.001$; Diff: 0.17; t: 3.65; P1–P3: $p < 0.001$; Diff: 0.20; t: 4.23
	m/min	64.6 ± 4.80	65.6 ± 3.88	116.2 ± 3.92	P1–P3: $p < 0.001$; Diff: −51.6; t: −8.30; P2–P3: $p < 0.001$; Diff: −50.5; t: −9.16

Regarding very high-speed distances, there were significant differences between P2 and P3 in both absolute and relative terms ($p < 0.001$). While P1 and P3 showed variations in relative volumes, no such differences were observed between P1 and P2 for either absolute or relative very high-speed running measurements.

4. Discussion

This study primarily aimed to clarify the ongoing debate concerning the efficacy of integrated versus decontextualized sprint-specific strategies on acceleration in professional women's football. Our findings suggest the potential for enhancing acceleration capabilities during competitive periods among trained players. Notably, both integrated and sprint-specific strategies, when emphasizing high-speed actions, exhibited positive impacts on sprinting ability across diverse distances.

In the context of team sports, professionals aim to balance player availability and performance [36]. Unlike individual sports, the relationship between fitness and performance in team sports is not straightforward [37]. This reality, compounded by dense competitive schedules, impacts loading strategies, often placing priority on player availability [38]. Notably, poor load management increases the risk of injury, particularly during periods of intense competition [39]. However, accepting a gradual decline or "plateau" in performance indicators such as acceleration over the course of a season can have negative effects on athletes. This is especially true since these conditions are not uniform among players throughout the season, so the stimulation to optimize such capabilities cannot be discarded.

Our results suggest that similar improvements in acceleration can be achieved using both sprint-specific methods and game-based approaches, provided the design of the game-based approaches allows for situations that necessitate high-intensity actions. Concerning resisted sprint training for female athletes, the literature still has a limited number of investigations on this topic. However, the results observed during this first period align closely with the findings of studies such as [40,41]. Both studies were of similar length, consisting of eight sessions, and despite the employment of different loads (12% BM and 40% BM), similar reductions in 30-m sprint times were reported. Specifically, reductions ranged from approximately 0.9 s for physically active women (T30: 5.14 s) to 0.16 s for highly trained handball players (T30: 4.81 s). In contrast, other studies focusing on analogous training methods have not provided detailed split times, choosing instead to highlight the impact on overall running velocity. Such studies observed minor enhancements during the later stages of acceleration without significant changes in the initial phases [42,43]. The variability in outcomes across these studies might stem from differences in intervention durations, participant characteristics, or specific loading dynamics, further complicating comparisons given the existing research gap [14,15].

Moreover, comparing the effectiveness of these methods with game-based strategies is challenging, mainly due to the lack of research focused on this objective. Most existing research on game-based interventions has demonstrated subpar outcomes in terms of enhancing acceleration when sport only is implemented [14], a finding reinforced by our study. Studies conducted by Djaoui et al. [44] or Kyprianou et al. [45] have examined game-based approaches, highlighting their inefficacy in reaching top speeds, which may be due to an overreliance on technical–tactical contexts or the intrinsic motivation of the athlete.

Performance improvements were noted in two conditions: (i) when specific sprint training was incorporated while maintaining a comparable overall workload to other intervals (e.g., P1 vs. P2); and (ii) when integrated training increased both absolute and relative workloads, particularly for key metrics (e.g., P1 vs. P3 and P2 vs. P3). In the periods characterized by integrated training that did not emphasize speed actions (P1 vs. P2), significant differences were only observed in variables such as "speed events" and "max dec". Both variables showed higher average intensities during P2. However, despite recording equal or lower average intensities for the previously mentioned variables, performance improvements were only notable when the stimuli of that period (P1) were complemented with resisted sprint training. In contrast, during P3, the increase in maximum deceleration intensity was notable. These higher deceleration intensities likely reflect increased overall physical demands, which may be partly due to improved acceleration capabilities or elevated velocities preceding decelerations as a result of task design modifications. For example, extending the acceleration phase before a change of direction—such as accelerating over 15 m compared to 5 m—results in higher approach speeds (approximately 83% versus 54% of maximum velocity), which, in turn, requires a greater number of braking steps (around five to six steps versus three steps) [46], and also increases the deceleration load not in terms of training volume but in intensity [47]. Specifically, T1 0–20 recorded 3.44 $\pm$ 0.03, T2 0–20 had 3.36 $\pm$ 0.04, and T3 0–20 was at 3.35 $\pm$ 0.03, with differences ΔT1–T2 = 0.08

and ΔT2–T3 = 0.01. Similarly, T1 0–30 had 4.81 ± 0.05, T2 0–30 was 4.66 ± 0.05, and T3 0–30 recorded 4.69 ± 0.04, with differences ΔT1–T2 = 0.15 and ΔT2–T3 = −0.03. Lastly, T1 20–30 showed 1.37 ± 0.01, T2 20–30 was 1.31 ± 0.01, and T3 20–30 registered 1.32 ± 0.02, with variations ΔT1–T2 = 0.05 and ΔT2–T3 = −0.01. During the final phase, P3 (T3–T4), notable reductions in sprint times were observed across all distances. Specifically, the 0–20 m segment showed a difference ΔT3–T4 = 0.06 ($p = 0.019$), the 0–30 m segment showed a difference ΔT3–T4 = 0.11 ($p < 0.001$), and the 20–30 m segment had a difference ΔT3–T4 = 0.05 ($p < 0.001$). These reductions indicate further improvements in acceleration capability, particularly during the latter portions of the sprint. It is noteworthy that longitudinal observational studies [48,49] have reported enhancements in acceleration performance during the preseason and first stages of competitive periods. Remarkably, these gains were observed even without specialized training interventions targeting these capabilities, suggesting that sport practice during these phases might inherently foster the development of such capabilities.

Nevertheless, reductions in high-intensity training volumes during the latter phase of the season, as observed by Mara [49], combined with the absence of specialized sprint training noted by Jiménez-Reyes et al. [48], have been associated with a decline in sprint performance, particularly during the initial phases of acceleration. Conversely, the ability to reach maximal speeds appears relatively consistent with limited inter-individual variation, as noted by López-Sagarra et al. [50].

Our findings emphasize the effectiveness of integrated training, especially when focused on high-speed activities, in enhancing acceleration capabilities across evaluated distances. Maintaining mean absolute and relative workloads through a game-centric integrated approach over an identical duration (6 weeks) was less effective in enhancing acceleration capabilities than when complemented with sprint-specific training (as evidenced by differences only in speed events and "max dec" between P1 and P3). However, the inclusion of high-speed metrics—such as distances covered at speeds exceeding 21 and 24 km/h, the number of speed events, or vigorous accelerations—led to significant performance improvements across all distances compared to integrated training alone.

Both individual studies [51,52] and systematic reviews [14,15] have found game-centric approaches like SSGs to be ineffective in enhancing sprinting capabilities. In the third phase of our study (P3), we prioritized strategies such as altering space [28], providing expansive play areas per player [25,53], and incorporating specific SSG variations termed 'transition games' [26,27,53] focused on counter-attacking scenarios with swift offensive and defensive transitions. Our results revealed improvements that surpassed those previously documented in the literature. While current evidence highlights the increased specificity of game-based approaches in elevating the frequency of high-intensity events, no studies have yet evaluated their long-term effects on in-season professional football players.

4.1. Limitations

Conducting research in elite athletic environments presents unique challenges. Throughout this study, the absence of players during evaluation sessions due to international duties was a notable limitation. Additionally, the dynamic nature of the team composition, influenced by transfer periods, further complicated our research. Given these challenges, we opted for an LMM approach, diverging from traditional analysis methods. This decision was driven by the difficulties in achieving a complete block design, underscoring the adaptability required in high-level research contexts. One limitation of the study is the extended time span between the second and third testing sessions, partly due to the inclusion of a transitional period. Although only the six weeks leading up to T3 were analyzed after the competitive regime resumed, and no significant workload differences were observed compared to the first intervention period, the variation in time intervals between testing sessions, particularly between T2 and T3, may have influenced the results.

4.2. Future Research Directions

While this study contributes valuable insights into the impact of integrated and sprint-specific training methods, further research is needed in several areas. First, more investigations are warranted on resisted sprint training in female athletes, as the current literature is still limited. Studies exploring optimal load configurations and their long-term impact on acceleration capabilities will help refine best practices. Additionally, further research on game-based approaches, particularly those incorporating specific high-intensity actions, is necessary to better understand their effects on sprint performance. Comparative studies examining the effectiveness of integrated strategies across various team sports and populations could provide a broader perspective on their utility.

4.3. Practical Implications

The findings of this study have important implications for the design of training programs in professional women's football. Coaches and performance staff should consider incorporating both sprint-specific and speed-oriented game-based training into their regimens, particularly during competitive periods. When game-based training is used, special attention should be given to designing drills that recreate high-speed events. Maintaining these high-intensity stimuli throughout the season is critical to avoiding declines in sprint performance. Each intervention period showed approximately a one-tenth of a second improvement in 0–20 m and 0–30 m sprints, equating to a 30–50 cm advantage in 30 m sprints during one-on-one duels, which can be decisive in match situations. These results offer valuable information supporting different strategies that can be incorporated into varied training contexts to enhance sprint performance.

5. Conclusions

Both high-speed game-based approaches and resisted sprint training have been identified as effective methods for enhancing acceleration capabilities. Resisted sprint training, in particular, delivers substantial benefits, even when high-intensity events are not specifically targeted during regular football training. Introducing training elements that increase high-intensity volumes prevents the decline of sprint performance over the competitive season, yielding improvements similar to those seen in sprint-specific interventions. In contrast, periods of training without a sprint focus led to changes in performance that were not statistically significant. These findings highlight the importance of incorporating structured sprint-focused elements into football training to maintain and improve acceleration throughout the season.

Author Contributions: Conceptualization, A.C.-Z., C.R. and P.J.-R.; Methodology, A.C.-Z.; Software, A.C.-Z.; Validation, A.C.-Z., C.R. and P.J.-R.; Formal Analysis, A.C.-Z.; Investigation, A.C.-Z., C.R. and P.J.-R.; Resources, A.C.-Z.; Data Curation, A.C.-Z.; Writing—Original Draft Preparation, A.C.-Z.; Writing—Review and Editing, A.C.-Z., C.R. and P.J.-R.; Visualization, A.C.-Z.; Supervision, P.J.-R.; Project Administration, A.C.-Z. and P.J.-R.; Funding Acquisition, A.C.-Z. and P.J.-R. All authors have read and agreed to the published version of the manuscript.

Funding: This study is part of the research project PID2019-108972RA-I00 named PROFIMATICS (PI: P.J.-R; team: 11 researchers; grant: EUR 37,000) funded by the "Ministerio de Ciencia e Innovación". This grant is in the framework of "Proyectos de I+D+i de los Programas Estatales de Generación de conocimiento y fortalecimiento científico y tecnológico del sistema I+D+i orientada a los retos de la sociedad". A.C.-Z. was granted a predoctoral contract "Contratos Predoctorales de Personal en Formación en Departamentos de la Universidad" at Rey Juan Carlos University, Spain.

Institutional Review Board Statement: This study protocol, in accordance with the principles of the Declaration of Helsinki, was granted approval by the research ethics committee at the Universidad Rey Juan Carlos (Ref: 0102202306623).

Informed Consent Statement: Informed consent was obtained from all subjects involved in the study.

Data Availability Statement: Publicly available data can be found at https://figshare.com/articles/dataset/Open_data/27079642 (accessed on 8 June 2024).

Acknowledgments: The authors thank all the athletes who participated as subjects in this study. Also, the authors would like to thank Valentin Romero for their technical support.

Conflicts of Interest: The authors declare no conflicts of interest.

References

1. Bradley, P.; Scott, D. Physical analysis of the FIFA Women's World Cup France 2019™. FIFA. 2019. Available online: https://digitalhub.fifa.com/m/4f40a98140d305e2/original/zijqly4oednqa5gffgaz-pdf.pdf (accessed on 8 June 2024).
2. Allen, T.; Taberner, M.; Zhilkin, M.; Rhodes, D. Running more than before? The evolution of running load demands in the English Premier League. *Int. J. Sports Sci. Coach.* **2023**, *19*, 779–787. [CrossRef]
3. Barnes, C.; Archer, D.T.; Hogg, B.; Bush, M.; Bradley, P.S. The evolution of physical and technical performance parameters in the english premier league. *Int. J. Sports Med.* **2014**, *35*, 1095–1100. [CrossRef] [PubMed]
4. Castillo, D.; Pérez-González, B.; Raya-González, J.; Fernández-Luna, Á.; Burillo, P.; Lago-Rodríguez, Á. Selection and promotion processes are not associated by the relative age effect in an elite Spanish soccer academy. *PLoS ONE* **2019**, *14*, e0219945. [CrossRef] [PubMed]
5. Dimundo, F.; Cole, M.; Blagrove, R.C.; McAuley, A.B.T.; Till, K.; Kelly, A.L. Talent Identification in an English Premiership Rugby Union Academy: Multidisciplinary Characteristics of Selected and Non-selected Male Under-15 Players. *Front. Sports Act. Living* **2021**, *3*, 688143. [CrossRef]
6. Nassis, G.P.; Massey, A.; Jacobsen, P.; Brito, J.; Randers, M.B.; Castagna, C.; Mohr, M.; Krustrup, P. Elite football of 2030 will not be the same as that of 2020: Preparing players, coaches, and support staff for the evolution. *Scand. J. Med. Sci. Sports* **2020**, *30*, 962–964. [CrossRef]
7. Faude, O.; Koch, T.; Meyer, T. Straight sprinting is the most frequent action in goal situations in professional football. *J. Sports Sci.* **2012**, *30*, 625–631. [CrossRef]
8. Glaise, P.; Rogowski, I.; Samozino, P.; Morin, J.B.; Morel, B.; Martin, C. Opposition Skill Efficiency During Professional Rugby Union Official Games Is Related to Horizontal Force-Production Capacities in Sprinting. *Int. J. Sports Physiol. Perform.* **2023**, *18*, 918–926. [CrossRef]
9. Heather, O.; Lander, P.; Rayner, R. Practice to pitch: The relationship between force-velocity profiles and match-day performance of semi-professional rugby union players. *Front. Sports Act. Living* **2023**, *5*, 1066767. [CrossRef]
10. Martínez-Hernández, D.; Quinn, M.; Jones, P. Most common movements preceding goal scoring situations in female professional soccer. *Sci. Med. Footb.* **2023**, *8*, 260–268. [CrossRef]
11. Haugen, T.A.; Breitschädel, F.; Seiler, S. Sprint mechanical properties in soccer players according to playing standard, position, age and sex. *J. Sports Sci.* **2020**, *38*, 1070–1076. [CrossRef]
12. Jiménez-Reyes, P.; Samozino, P.; García-Ramos, A.; Cuadrado-Peñafiel, V.; Brughelli, M.; Morin, J.B. Relationship between vertical and horizontal force-velocity-power profiles in various sports and levels of practice. *PeerJ* **2018**, *6*, e5937. [CrossRef] [PubMed] [PubMed Central]
13. Bolger, R.; Lyons, M.; Harrison, A.J.; Kenny, I.C. Sprinting performance and resistance-based training interventions: A systematic review. *J. Strength. Cond. Res.* **2015**, *29*, 1146–1156. [CrossRef] [PubMed]
14. Nicholson, B.; Dinsdale, A.; Jones, B.; Till, K. The Training of Short Distance Sprint Performance in Football Code Athletes: A Systematic Review and Meta-Analysis. *Sports Med.* **2021**, *51*, 1179–1207. [CrossRef] [PubMed]
15. Nicholson, B.; Dinsdale, A.; Jones, B.; Till, K. The Training of Medium- to Long-Distance Sprint Performance in Football Code Athletes: A Systematic Review and Meta-analysis. *Sports Med.* **2022**, *52*, 257–286. [CrossRef]
16. Rumpf, M.C.; Lockie, R.G.; Cronin, J.B.; Jalilvand, F. Effect of Different Sprint Training Methods on Sprint Performance over Various Distances: A Brief Review. *J. Strength Cond. Res.* **2016**, *30*, 1767–1785. [CrossRef]
17. Haugen, T. Sprint conditioning of elite soccer players: Worth the effort or let's just buy faster players? *Sport. Perform. Sci. Rep.* **2017**, *1*, 1–2.
18. Alcaraz, P.E.; Carlos-Vivas, J.; Oponjuru, B.O.; Martínez-Rodríguez, A. The Effectiveness of Resisted Sled Training (RST) for Sprint Performance: A Systematic Review and Meta-analysis. *Sports Med.* **2018**, *48*, 2143–2165. [CrossRef]
19. Cross, M.R.; Lahti, J.; Brown, S.R.; Chedati, M.; Jimenez-Reyes, P.; Samozino, P.; Eriksrud, O.; Morin, J.B. Training at maximal power in resisted sprinting: Optimal load determination methodology and pilot results in team sport athletes. *PLoS ONE* **2018**, *13*, e0195477. Available online: https://pubmed.ncbi.nlm.nih.gov/29641589/ (accessed on 6 November 2023). [CrossRef]
20. Lahti, J.; Jiménez-Reyes, P.; Cross, M.R.; Samozino, P.; Chassaing, P.; Simond-Cote, B.; Ahtiainen, J.P.; Morin, J.B. Individual Sprint Force-Velocity Profile Adaptations to In-Season Assisted and Resisted Velocity-Based Training in Professional Rugby. *Sports* **2020**, *8*, 74. Available online: https://pubmed.ncbi.nlm.nih.gov/32466235/ (accessed on 29 January 2024). [CrossRef]
21. Morin, J.B.; Capelo-Ramirez, F.; Rodriguez-Pérez, M.A.; Cross, M.R.; Jimenez-Reyes, P. Individual Adaptation Kinetics Following Heavy Resisted Sprint Training. *J. Strength Cond. Res.* **2022**, *36*, 1158–1161. Available online: https://pubmed.ncbi.nlm.nih.gov/32058358/ (accessed on 29 January 2024). [CrossRef]

22. Douchet, T.; Paizis, C.; Carling, C.; Cometti, C.; Babault, N. Typical weekly physical periodization in French academy soccer teams: A survey. *Biol. Sport.* **2023**, *40*, 731–740. [CrossRef] [PubMed]
23. Dios-Álvarez, V.; de Castellano, J.; Padrón-Cabo, A.; Rey, E. Do small-sided games prepare players for the worst-case scenarios of match play in elite young soccer players? *Biol. Sport* **2024**, *41*, 95–106. [CrossRef] [PubMed]
24. Martín-García, A.; Castellano, J.; Méndez Villanueva, A.; Gómez-Díaz, A.; Cos, F.; Casamichana, D. Demands of Ball Possession Games in Relation to the Most Demanding Passages of a Competitive Match. Available online: http://www.jssm.org (accessed on 8 June 2024).
25. Riboli, A.; Coratella, G.; Rampichini, S.; Ce, E.; Esposito, F. Area per player in small-sided games to replicate the external load and estimated physiological match demands in elite soccer players. *PLoS ONE* **2020**, *15*, e0229194. [CrossRef] [PubMed]
26. Bortnik, L.; Burger, J.; Morgans, R.; Rhodes, D. Utilisation of transitional clusters exhibited within soccer game play to inform training design. *Sci. J. Sport Perform.* **2023**, *2*, 439–453. [CrossRef]
27. Bortnik, L.; Burger, J.; Rhodes, D. The mean and peak physical demands during transitional play and high-pressure activities in elite football. *Biol. Sport.* **2022**, *39*, 1055–1064. [CrossRef]
28. Asian-Clemente, J.A.; Rabano-Muñoz, A.; Suarez-Arrones, L.; Requena, B. Different pitch configurations constrain the external and internal loads of young professional soccer players during transition games. *Biol. Sport.* **2023**, *40*, 1047–1055. [CrossRef]
29. Bramah, C.; Mendiguchia, J.; Dos'Santos, T.; Morin, J.B. Exploring the Role of Sprint Biomechanics in Hamstring Strain Injuries: A Current Opinion on Existing Concepts and Evidence. *Sports Med.* **2023**, *54*, 783–793. Available online: https://link.springer.com/article/10.1007/s40279-023-01925-x (accessed on 29 January 2024). [CrossRef]
30. Brughelli, M.; Cronin, J.; Mendiguchia, J.; Kinsella, D.; Nosaka, K. Contralateral leg deficits in kinetic and kinematic variables during running in Australian Rules football players with previous hamstring injuries. *J. Strength. Cond. Res.* **2010**, *24*, 2539–2544. [CrossRef]
31. Edouard, P.; Mendiguchia, J.; Guex, K.; Lahti, J.; Samozino, P.; Morin, J.B. Sprinting: A Potential Vaccine for Hamstring Injury? Available online: https://sportperfsci.com/wp-content/uploads/2019/01/SPSR55_Edouard_190108_final.pdf (accessed on 15 June 2024).
32. Mendiguchia, J.; Samozino, P.; Brughelli, M.; Schmikli, S.; Morin, J.B.; Mendez-Villanueva, A. Progression of Mechanical Properties during On- fi eld Sprint Running after Returning to Sports from a Hamstring Muscle Injury in Soccer Players. *Int. J. Sports Med.* **2014**, *35*, 690–695.
33. Romero, V.; Lahti, J.; Castaño Zambudio, A.; Mendiguchia, J.; Jiménez Reyes, P.; Morin, J.B. Effects of Fatigue Induced by Repeated Sprints on Sprint Biomechanics in Football Players: Should We Look at the Group or the Individual? *Int. J. Environ. Res. Public Health* **2022**, *19*, 14643. Available online: https://pubmed.ncbi.nlm.nih.gov/36429363/ (accessed on 5 February 2024). [CrossRef]
34. Mendiguchia, J.; Edouard, P.; Samozino, P.; Brughelli, M.; Cross, M.; Ross, A.; Gill, N.; Morin, J.B. Field monitoring of sprinting power–force–velocity profile before, during and after hamstring injury: Two case reports. *J. Sports Sci.* **2016**, *34*, 535–541. [CrossRef] [PubMed]
35. Newans, T.; Bellinger, P.; Drovandi, C.; Buxton, S.; Minahan, C. The Utility of Mixed Models in Sport Science: A Call for Further Adoption in Longitudinal Data Sets. *Int. J. Sports Physiol. Perform.* **2022**, *17*, 1289–1295. [CrossRef] [PubMed]
36. Gabbett, T.J. The training-injury prevention paradox: Should athletes be training smarter and harder? *Br. J. Sports Med.* **2016**, *50*, 273–280. [CrossRef] [PubMed]
37. Boullosa, D.; Casado, A.; Claudino, J.G.; Jiménez-Reyes, P.; Ravé, G.; Castaño-Zambudio, A.; Lima-Alves, A.; de Oliveira, S.A., Jr.; Dupont, G.; Granacher, U.; et al. Do you Play or Do you Train? Insights From Individual Sports for Training Load and Injury Risk Management in Team Sports Based on Individualization. *Front Physiol.* **2020**, *11*, 995. [CrossRef]
38. Calleja-González, J.; Mallo, J.; Cos, F.; Sampaio, J.; Jones, M.T.; Marqués-Jiménez, D.; Mielgo-Ayuso, J.; Freitas, T.T.; Alcaraz, P.E.; Vilamitjana, J.; et al. A commentary of factors related to player availability and its influence on performance in elite team sports. *Front. Sports Act. Living* **2023**, *4*, 1077934. [CrossRef]
39. Page, R.M.; Field, A.; Langley, B.; Harper, L.D.; Julian, R. The Effects of Fixture Congestion on Injury in Professional Male Soccer: A Systematic Review. *Sports Med.* **2023**, *53*, 667–685. [CrossRef]
40. Luteberget, L.S.; Raastad, T.; Seynnes, O.; Spencer, M. Effect of traditional and resisted sprint training in highly trained female team handball players. *Int. J. Sports Physiol. Perform.* **2015**, *10*, 642–647. [CrossRef]
41. Pareja-Blanco, F.; de Villarreal, E.S.; Bachero-Mena, B.; Mora-Custodio, R.; Asián-Clemente, J.A.; Loturco, I.; Rodríguez-Rosell, D. Effects of unloaded sprint and heavy sled training on sprint performance in physically active women. *Int. J. Sports Physiol. Perform.* **2020**, *15*, 1356–1362. [CrossRef]
42. Makaruk, B.; Sozański, H.; Makaruk, H.; Sacewicz, T. The effects of resisted sprint training on speed performance in women. *Hum. Mov.* **2013**, *14*, 116–122. [CrossRef]
43. Upton, D.E. The Effect of Assisted and Resisted Sprint Training on Acceleration and Velocity in Division Ia Female Soccer Athletes. Available online: www.nsca-jscr.org (accessed on 8 June 2024).
44. Djaoui, L.; Chamari, K.; Owen, A.L.; Dellal, A. Maximal sprinting speed of elite soccer players during training and matches. *J. Strength Cond. Res.* **2017**, *31*, 1509–1517. [CrossRef]
45. Kyprianou, E.; Di Salvo, V.; Lolli, L.; Al Haddad, H.; Villanueva, A.M.; Gregson, W.; Weston, M. To Measure Peak Velocity in Soccer, Let the Players Sprint. *J. Strength Cond. Res.* **2019**, *36*, 273–276. [CrossRef] [PubMed]

46. Harper, D.; Cervantes, C.; Van Dyke, M.; Evans, M.; McBurnie, A.; Dos' Santos, T.; Eriksrud, O.; Cohen, D.; Rhodes, D.; Carling, C.; et al. The Braking Performance Framework: Practical Recommendations and Guidelines to Enhance Horizontal Deceleration Ability in Multi-Directional Sports. *Int. J. Strength Cond.* **2024**, *4*. Available online: https://journal.iusca.org/index.php/Journal/article/view/351 (accessed on 16 October 2024).
47. Gaudino, P.; Alberti, G.; Iaia, F.M. Estimated metabolic and mechanical demands during different small-sided games in elite soccer players. *Hum. Mov. Sci.* **2014**, *36*, 123–133. [CrossRef] [PubMed]
48. Jiménez-Reyes, P.; Garcia-Ramos, A.; Párraga-Montilla, J.A.; Morcillo-Losa, J.A.; Cuadrado-Peñafiel, V.; Castaño-Zambudio, A.; Samozino, P.; Morin, J.B. Seasonal Changes in the Sprint Acceleration Force-Velocity Profile of Elite Male Soccer Players. *J. Strength Cond. Res.* **2022**, *36*, 70–74. Available online: https://pubmed.ncbi.nlm.nih.gov/32329976/ (accessed on 11 October 2022). [CrossRef] [PubMed]
49. Mara, J.K.; Thompson, K.G.; Pumpa, K.L.; Ball, N.B. Periodization and physical performance in elite female soccer players. *Int. J. Sports Physiol. Perform.* **2015**, *10*, 664–669. [CrossRef]
50. López-Sagarra, A.; Baena-Raya, A.; Casimiro-Artés, M.; Granero-Gil, P.; Rodríguez-Pérez, M.A. Seasonal Changes in the Acceleration–Speed Profile of Elite Soccer Players: A Longitudinal Study. *Appl. Sci.* **2022**, *12*, 12987. Available online: https://www.mdpi.com/2076-3417/12/24/12987/htm (accessed on 20 March 2023). [CrossRef]
51. Derakhti, M.; Bremec, D.; Kambic, T.; Siethoff LTen Psilander, N. Four Weeks of Power Optimized Sprint Training Improves Sprint Performance in Adolescent Soccer Players. *Int. J. Sports Physiol. Perform.* **2022**, *17*, 1343–1351. [CrossRef]
52. Mendiguchia, J.; Conceição, F.; Edouard, P.; Fonseca, M.; Pereira, R.; Lopes, H.I.; Morin, J.B.; Jiménez-Reyes, P. Sprint versus isolated eccentric training: Comparative effects on hamstring architecture and performance in soccer players. *PLoS ONE* **2020**, *15*, e0228283. [CrossRef]
53. Asian-Clemente, J.; Rabano-Muñoz, A.; Muñoz, B.; Franco, J.; Suarez-Arrones, L. Can Small-side Games Provide Adequate High-speed Training in Professional Soccer? *Int. J. Sports Med.* **2021**, *42*, 523–528. [CrossRef]

applied sciences

MDPI

Article

Trend Forecasting in Swimming World Records and in the Age of World Record Holders

Mário J. Costa [1,*], Luis Quinta-Nova [2], Sandra Ferreira [3], Aldo M. Costa [4,5] and Catarina C. Santos [1,6,7]

[1] Centre of Research, Education, Innovation, and Intervention in Sport (CIFI2D), and Porto Biomechanics Laboratory (LABIOMEP-UP), Faculty of Sport, University of Porto, 4050-4050 Porto, Portugal; cmsantos@fade.up.pt
[2] Higher Agricultural School, Polytechnic University of Castelo Branco, 6000-084 Castelo Branco, Portugal; lnova@ipcb.pt
[3] Department of Mathematics and Centre of Mathematics and Applications, University of Beira Interior, 6201-001 Covilhã, Portugal; sandraf@ubi.pt
[4] Department of Sport Sciences, University of Beira Interior, 6201-001 Covilhã, Portugal; amcosta@ubi.pt
[5] Research Center in Sport Sciences, Health Sciences and Human Development (CIDESD), 6201-001 Covilhã, Portugal
[6] Higher Education School, Polytechnic of Coimbra, 3030-329 Coimbra, Portugal
[7] Department of Sport Sciences, Higher Institute of Educational Sciences of the Douro (ISCE-Douro), 4560-547 Penafiel, Portugal
* Correspondence: mjcosta@fade.up.pt

Abstract: This study aimed to forecast trends in swimming world records (WRs) and in the age of record holders. A total of 566 individual freestyle WRs (290 for males and 276 for females) were retrieved from open access websites. The frequency of observations in WRs in each decade and event was computed for males and females. The swimmers' chronological age was converted into decimal age at the time of breaking the world record. ARIMA forecasting models and exponential smoothing techniques were used to examine historical trends and predict future observations. The WRs improved over time, and there was a nuanced pattern in the age of world record holders. While certain events (50 m and 100 m) showed swimmers achieving records at older ages, others (e.g., 200 m, 400 m, 800 m, and 1500 m) displayed variations. Forecasting shows a continuing improvement in WRs in the upcoming years, with the age of male world record holders stabilizing in shorter events and decreasing in longer distance ones, while for females, general stabilization should be expected for the majority of competitive events.

Keywords: swimmers; performance; decades; prediction; freestyle

1. Introduction

Success in competitive swimming requires a clearly defined pathway of progression to deliver the best performances in adulthood. In the past, several backcourt strategies have been implemented to help swimmers have a longer and more sustainable career. The implementation of Long-Term Athletic Development models [1,2] or performance monitoring over swimmers' careers [3–5] contributed to understanding how swimmers behave over the years and how they may improve in the long-term. Moreover, the determination of normative data and percentile curves [6] and the comparison of age-related performance progression against top athletes [7] have also been used for tracking swimmers headed toward future success.

At some point in a swimmer's career, there will be stabilization in their performance near the peak. The age range between 15 and 16 years has been considered a determinant for achieving the best individual performance in long-distance events [8]. Males' peak individual performance seems to occur between 18 and 19 years for short-distance specialists and between 17 and 18 years for long-distance specialists [9]. The analysis of overall trends and the individual paths of performance over a decade showed deceleration around the age of 16 years [4], with the peak performance being reached around 18–19 years for young adults [10].

World-class performances are also highly dependent on swimmers' sex. Allen et al. [7] concluded that top male swimmers achieved peak performance ~2 years later than their female counterparts. This difference in age of peak performance between sexes appears to be explained almost entirely by the ~2-year earlier onset of puberty in females compared with males [11]. However, during recent decades, the improved performance of elite female athletes has narrowed the gap by 8 to 12% compared to male athletes [12]. Despite that, some genetic traits related to athletic development and peak performance will always persist [13], not allowing for the complete disappearance of this gender gap.

As swimmers grow older, they may become unwilling or physiologically unable to sustain the training loads required to achieve top performances in their events. Although these swimmers may endure the training loads, the continuous training stress may increase their susceptibility to injury [14]. This scenario could result in a premature peak in performance, leading to a career break or a drop-out. As a consequence, most of these decisions are taken to pursue other challenges like parenting, pursuing an academic/professional career, or even just recovering from the emotional or physical exhaustion induced by training [15]. Still, if they return to practice a few years later, they can compete at an elite level (e.g., Dara Torres, Rūta Meilutytė, or Cody Simpson).

Recent top-level performances in shorter distances (including some in World and Olympic events) were achieved by swimmers who were over 30 years old (e.g., Roland Schoeman, Jason Lezak, Marleen Veldhuis, Nicholas Santos, Sarah Sjöström). It seems that some of today's top-ranked swimmers are winning at a more advanced age than their previous counterparts. Knetchle et al. [16] found that the age of peak performance of World and Olympic champions competing between 1992 and 2013 was ~2–3 years older compared with Olympic gold medal winners (aged ~21 and ~23 years) from 1896 to 1980. Rust et al. [17] also found that the age of peak swimming speed for elite female swimmers increased in the 50 m freestyle from 1994 to 2012.

World records in long-course freestyle swimming have constantly improved over the decades, influenced by changes in training practices and/or technology. However, it remains unclear if there is a preference in the age of world record holders across different types of freestyle events (short-distance vs. long-distance). Studying world record dynamics based on historical data can provide insights into world record trends and whether they differ by sex and event type. As athletes continually strive for excellence, it is intriguing to examine whether there is a shift in the age at which swimmers achieve those world records. To the best of the authors' knowledge, there is a lack of research on world records in swimming, as no research has been conducted in this context. So, there is a chance here to conduct a deeper analysis of world record trends and predict what can happen in the years to come.

This study aimed to achieve the following: (i) perform a comprehensive analysis of historical data on swimming world records; (ii) identify patterns and trends in world records and in the age of record-breaking athletes; and (iii) predict the next two world records and the age of the world record holders. The following has been hypothesized: (i) there is a trend of improvement in world records over time in long-course events; (ii) the age of world record holders differs between short-distance and long-distance freestyle events; (iii) future world records might be accurately predicted based on historical data, with discrepancies in trends based on sex and event type.

2. Materials and Methods

2.1. Design

An observational retrospective and prospective design was selected for the present study. A comprehensive dataset of swimmers' demographics and world records (WRs) in freestyle swimming events was compiled by two independent researchers consulting open access websites (e.g., swimrankings.net; omegatiming.com; worldaquatics.com). The time frame of this study covered 10 decades to capture long-term trends. As such, the WRs in 50 m, 100 m, 200 m, 400 m, 800 m, and 1500 m freestyle events (long-course) from 1920 up to 2019 were considered. The event, swimmer's name, chronological age, and date of the record were extracted for each swimming event and grouped by decade (e.g., 1920–1929). The chronological age of each swimmer was converted into decimal age at the time of breaking the world record. A total of 566 individual WRs (290 for males and 276 for females) were included for further analysis. All records have been converted from min to s if applicable to the event. The Institutional Ethics Committee stated that ethics approval was not required for this type of study design.

2.2. Statistical Analysis

The frequency of observations in WRs in each decade and event was computed. The Kolmogorov–Smirnov test was used to analyze the normality assumption. The mean plus one standard deviation was used for descriptive statistics and divided according to sex. Since the aim was to detect overall temporal trends across decades, rather than specific interdecadal comparisons, a one-way ANOVA was used to assess differences in WRs and in the age of world record holders between decades. The Levene test was used to evaluate the homogeneity assumption. The effect size was collected through Eta Squared (η^2) and interpreted according to Ferguson as follows [18]: no effect if $\eta^2 \leq 0.04$, a minimal effect if $0.04 > \eta^2 \leq 0.25$, a moderate effect if $0.25 > \eta^2 \leq 0.64$, and a strong effect if $\eta^2 > 0.64$).

Autoregressive Integrated Moving Average (ARIMA) models were used to examine historical trends and predict future observations among WRs and the age of world record holders. For this purpose, a univariate time series was used with a sequence of measurements of the same variable collected over time. For each model, the forecast started after the last non-missing value in the range of the requested estimation period and ended at the last period for which non-missing values of all of the predictors were available or at the end date of the requested forecast period, whichever was earlier. This approach allowed each event to be described in terms of time and/or age as either increasing, decreasing, or stabilizing. Just two observations were considered for the description of the forecast trends within the next decade.

All statistical analyses were performed using SPSS software (v.27, IBM, SPSS Inc., Chicago, IL, USA), with statistical significance set at $p \leq 0.05$.

3. Results

Table 1 shows the frequency of WRs in freestyle events over the decades according to sex. While there was a trend for a higher frequency of WRs in both groups between the 1960s and the beginning of the 21st century, it has slowed down in the last decade.

Table 1. The frequency of WRs in freestyle events (long-course pool) over the decades according to sex.

	Decades									
	1920–1929	1930–1939	1940–1949	1950–1959	1960–1969	1970–1979	1980–1989	1990–1999	2000–2009	2010–2019
Males										
50 m	--	--	--	--	--	3	15	2	7	--
100 m	3	3	3	4	7	9	5	1	8	--
200 m	4	1	2	9	17	9	8	3	8	--
400 m	5	3	4	10	13	14	7	4	5	--
800 m	6	7	5	5	6	13	3	3	4	--
1500 m	6	--	2	4	9	11	3	4	1	2
Females										
50 m	--	--	--	--	--	2	13	2	8	1
100 m	7	6	10	5	13	3	2	11	2	--
200 m	3	6	--	5	7	14	2	1	8	--
400 m	7	9	2	2	13	12	2	--	6	3
800 m	--	4	1	6	9	19	4	--	1	5
1500 m	4	3	2	1	10	11	2	--	1	6

Table 2 shows the differences between decades in mean WRs according to sex. There was a natural trend of improvement in WRs over the decades, with a strong effect in all of the events analyzed.

Table 2. Differences between decades in mean WR times (in s) according to sex.

	Decades										One-Way ANOVA		
	1920–1929	1930–1939	1940–1949	1950–1959	1960–1969	1970–1979	1980–1989	1990–1999	2000–2009	2010–2019	F	p	η^2
Males													
50 m	--	--	--	--	--	23.77 ± 0.08	22.56 ± 0.43	21.90 ± 0.12	21.32 ± 0.36	--	34.44	<0.01	0.82
100 m	58.80 ± 1.51	56.60 ± 0.20	55.70 ± 0.26	55.00 ± 0.37	53.03 ± 0.74	50.81 ± 0.78	48.94 ± 1.37	48.21 *	47.44 ± 0.43	--	147.71	<0.01	0.97
200 m	134.65 ± 4.90	127.20 *	125.80 ± 0.57	123.12 ± 1.51	117.85 ± 2.35	111.65 ± 1.53	107.90 ± 0.85	106.34 ± 0.34	104.35 ± 1.36	--	132.75	<0.01	0.95
400 m	300.08 ± 8.86	284.03 ± 4.63	275.15 ± 2.38	265.24 ± 4.88	250.42 ± 3.15	236.12 ± 4.52	228.82 ± 1.63	224.28 ± 1.96	220.45 ± 0.54	--	235.80	<0.01	0.97
800 m	642.03 ± 15.17	610.66 ± 9.10	583.54 ± 5.77	555.64 ± 11.06	521.80 ± 8.67	492.53 ± 8.99	471.90 ± 1.13	466.82 ± 0.94	457.88 ± 4.05	--	287.54	<0.01	0.98
1500 m	1233.32 ± 56.56	--	1107.35 ± 11.81	1071.75 ± 16.26	1001.54 ± 24.60	930.24 ± 18.96	896.46 ± 1.76	885.98 ± 4.08	874.56 *	872.58 ± 2.21	90.23	<0.01	0.96
Females													
50 m	--	--	--	--	--	26.87 ± 0.18	25.67 ± 0.44	24.65 ± 0.20	24.16 ± 0.28	23.67 *	35.76	<0.01	0.87
100 m	71.74 ± 2.01	65.90 ± 1.27	62.67 ± 1.16	59.70 ± 0.51	57.14 ± 1.17	54.83 ± 0.13	54.25 ± 0.83	53.10 ± 0.62	51.89 ± 0.25	--	204.61	<0.01	0.97
200 m	164.47 ± 3.56	147.07 ± 4.41	--	138.18 ± 2.24	129.24 ± 1.63	121.84 ± 2.92	117.65 ± 0.14	116.78 *	115.00 ± 1.28	--	170.99	<0.01	0.97
400 m	354.25 ± 18.72	315.71 ± 8.55	302.75 ± 3.75	289.00 ± 2.55	276.64 ± 6.65	255.01 ± 6.13	244.65 ± 1.13	--	241.15 ± 1.37	237.90 ± 1.27	126.42	<0.01	0.96
800 m	--	702.23 ± 27.89	652.50 *	621.73 ± 12.36	573.03 ± 16.29	525.90 ± 12.09	498.83 ± 2.78	--	494.10 *	488.57 ± 3.35	133.26	<0.01	0.96
1500 m	1454.75 ± 35.87	1353.20 ± 45.85	1263.55 ± 9.26	1246.50 *	1001.28 ± 42.24	997.49 ± 20.14	956.42 ± 6.10	--	942.54 *	928.80 ± 5.85	153.53	<0.01	0.98

* Single WR.

Table 3 shows the differences between decades in the mean age of world record holders according to sex. Preliminary data analysis reveals a nuanced pattern in the age of world record holders. While certain events (e.g., 50 m and 100 m) show a consistent trend of swimmers achieving records at older ages over the decades, other events (e.g., 200 m, 400 m, 800 m, and 1500 m) display variations.

Table 3. The differences between decades in the mean age (decimal age) of WR holders according to sex.

	Decades										One-Way ANOVA		
	1920–1929	1930–1939	1940–1949	1950–1959	1960–1969	1970–1979	1980–1989	1990–1999	2000–2009	2010–2019	F	p	η^2
Males													
50 m	--	--	--	--	--	23.44 ± 1.88	21.84 ± 2.42	25.46 ± 0.00	24.58 ± 2.69	--	2.92	0.05	0.28
100 m	22.61 ± 6.45	21.20 ± 0.97	20.18 ± 1.41	22.26 ± 2.64	21.06 ± 1.80	21.05 ± 1.62	21.07 ± 1.37	22.59 *	23.64 ± 1.17	--	1.30	0.28	--
200 m	21.23 ± 2.24	20.52 *	18.72 ± 1.44	19.26 ± 2.55	20.08 ± 2.21	19.58 ± 1.99	20.48 ± 1.38	17.53 ± 1.16	20.85 ± 2.44	--	1.16	0.34	--
400 m	20.97 ± 2.67	19.98 ± 1.76	19.36 ± 1.89	18.76 ± 2.03	19.24 ± 2.12	18.01 ± 0.94	20.78 ± 1.34	19.02 ± 1.76	19.41 ± 2.17	--	2.15	0.05	0.24
800 m	20.18 ± 3.66	19.70 ± 2.35	20.14 ± 1.65	17.57 ± 2.37	21.11 ± 1.94	16.91 ± 1.07	23.58 ± 2.24	19.19 ± 1.61	21.25 ± 3.23	--	4.75	<0.01	0.47
1500 m	22.53 ± 3.17	--	20.91 ± 0.01	18.94 ± 2.63	20.45 ± 2.33	17.57 ± 2.69	21.49 ± 1.31	19.90 ± 1.27	21.22 *	20.17 ± 0.72	2.52	0.03	0.38
Females													
50 m	--	--	--	--	--	17.91 ± 1.64	17.02 ± 1.38	23.65 ± 4.38	25.94 ± 3.27	23.95 *	19.27	<0.01	0.79
100 m	18.48 ± 1.01	16.78 ± 1.02	19.12 ± 1.20	22.48 ± 4.65	16.98 ± 3.38	20.87 ± 0.29	19.25 ± 0.34	23.93 ± 2.71	24.11 ± 0.25	--	8.54	<0.01	0.58
200 m	18.52 ± 3.04	16.48 ± 1.14	--	19.07 ± 1.28	17.41 ± 2.55	16.28 ± 1.39	17.23 ± 1.50	16.42 *	20.75 ± 1.63	--	5.87	<0.01	0.52
400 m	18.15 ± 1.45	17.22 ± 0.86	19.75 ± 0.01	17.98 ± 0.11	16.67 ± 1.05	16.54 ± 1.14	16.69 ± 0.54	--	20.57 ± 1.13	18.07 ± 1.14	9.86	<0.01	0.63
800 m	--	21.15 ± 3.82	20.67 *	15.72 ± 2.56	15.79 ± 1.44	16.26 ± 1.49	17.11 ± 1.04	--	19.49 *	18.06 ± 1.22	5.22	<0.01	0.47
1500 m	18.52 ± 1.86	18.03 ± 0.45	20.18 ± 0.72	16.27 *	15.47 ± 1.06	15.10 ± 1.20	16. 25 ± 0.45	--	18.97 *	18.16 ± 1.65	8.14	<0.01	0.68

* Single WR holder.

Figures 1 and 2 show the forecast models for the next two WRs in male and female freestyle events, respectively. The future observations show a natural improvement in WRs for both sexes in the upcoming decades.

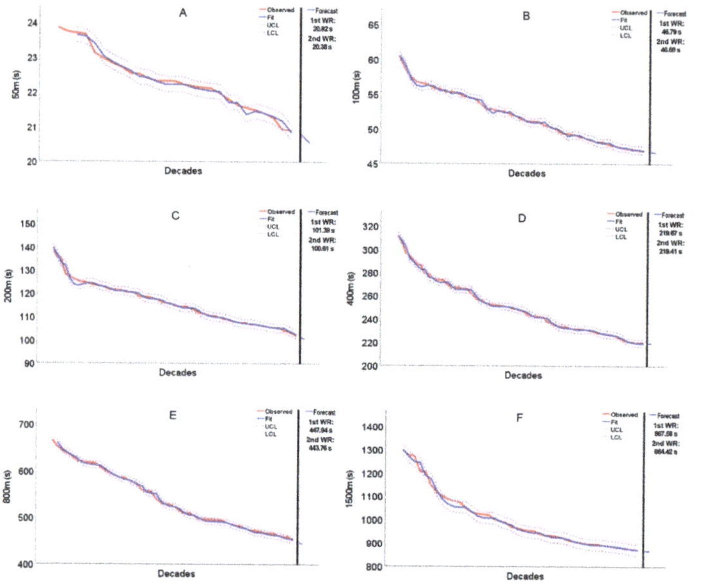

Figure 1. Forecast models for predicting male WRs in all freestyle events (panel (**A**), 50 m; panel (**B**), 100 m; panel (**C**), 200 m; panel (**D**), 400 m; panel (**E**), 800 m; panel (**F**), 1500 m). Fit: line of best fit; UCL: upper confidence limit; LCL: lower confidence limit; WR: predicted time for world record.

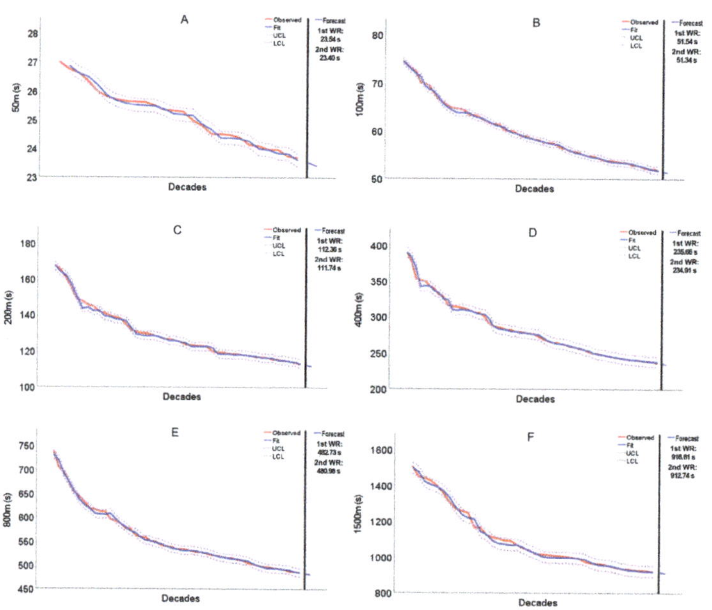

Figure 2. Forecast models for predicting female WRs in all freestyle events (panel (**A**), 50 m; panel (**B**), 100 m; panel (**C**), 200 m; panel (**D**), 400 m; panel (**E**), 800 m; panel (**F**), 1500 m). Fit: line of best fit; UCL: upper confidence limit; LCL: lower confidence limit; WR: predicted time for world record.

Figures 3 and 4 show the forecast models for the next two ages of male and female world record holders, respectively. The future observations show that the age of male world record holders seems to stabilize in shorter events and decrease in longer distance ones, while for females, general stabilization should be expected for the majority of the events.

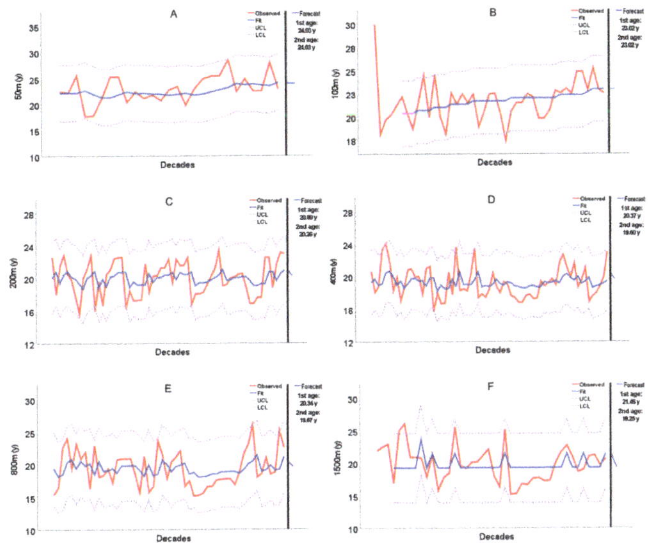

Figure 3. Forecast models for predicting age of male world record holders in all freestyle events (panel (**A**), 50 m; panel (**B**), 100 m; panel (**C**), 200 m; panel (**D**), 400 m; panel (**E**), 800 m; panel (**F**), 1500 m). Fit: line of best fit; UCL: upper confidence limit; LCL: lower confidence limit.

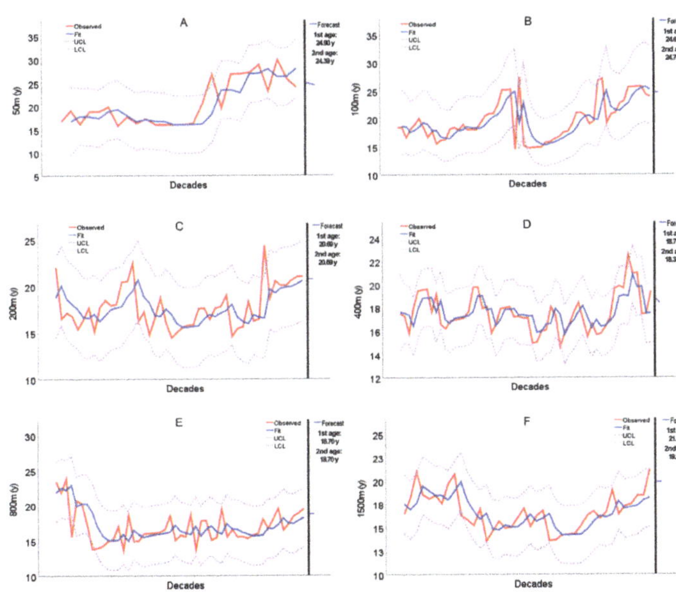

Figure 4. Forecast models for predicting age of female world record holders in freestyle events (panel (**A**), 50 m; panel (**B**), 100 m; panel (**C**), 200 m; panel (**D**), 400 m; panel (**E**), 800 m; panel (**F**), 1500 m). Fit: line of best fit; UCL: upper confidence limit; LCL: lower confidence limit.

4. Discussion

This study aimed to perform an analysis of historical data on swimming world records and the age of world record holders to identify trends and predict future events. There was a natural trend of improvement in WRs over the decades, with world record holders becoming older mainly in short-distance events. The future observations show a continuing improvement in WRs, with general stabilization in the age of world record holders but with a slight decrease in the age of males in long-distance events.

The retrospective analysis of WR evolution revealed a natural trend of improvement over the decades, consistent with broader trends seen in other sports [19]. The advancements in science and technology, such as better swimwear, training equipment, and monitoring devices, have enabled athletes to push the boundaries of their capabilities. Beyond technological support, swimming has become more multidisciplinary with technical teams being more complete, higher in number, and more specialized in the different aspects of performance. Over the years, swimmers have had access to more sophisticated training programs, personalized nutrition or recovery plans, and in-depth analysis of racing strategies [20], all of which contributed to enhanced performance and record-breaking achievements.

The growth of swimming as a sport has also led to increased participation and competition at both the elite and grassroots levels (e.g., as seen in the Michael Phelps effect). Greater competition at younger ages often drives athletes to train harder and strive for excellence, leading to improvements in performance at an adult age [21]. The changes in the sport itself by the periodic amendments to the rules and regulations governing swimming competitions may also have influenced WR evolution. For example, changes in pool design, including deeper, deck-level pools, and more effective anti-wave lane ropes [22] and starting blocks [23], have impacted swimming performances and contributed to the progression of WRs over time. These changes have made the sport more complete and up-to-date and have triggered the globalization of swimming to another level, exposing athletes to different coaching philosophies and competition styles from around the world. This exchange of knowledge and expertise drives innovation and improvement in swimming performance on a global scale.

The analysis of WRs revealed a notable trend in record frequency over the past few decades. Specifically, it was observed that between 1960 and 2010, there was a consistent increase in the frequency of world records being broken, which is in agreement with previous studies [24]. It is plausible that the significant improvements in swimming technology, evaluation techniques, and training processes during the mid- to late 20th century, as well as the scientific development in swimming suit design up to 2010, led to more rapid advancements and thus a higher frequency of record breaking. The beginning of the 21st century, mainly between the 2004 Athens Olympic Games and the 2009 Rome World Championships, represented the "shiny-suit era", with a large number of WRs being broken. However, it has been over a decade since FINA (nowadays World Aquatics) banned full-body tech suits, which can help explain the slowdown in the rate of world record breaking in the last couple of years.

Contrary to this trend in WRs, there was a nuanced pattern observed in the age of world record holders over the decades. Previous analyses of Olympic swimming data showed that the age of peak performance has remained remarkably consistent until later in the 1980s [25]. However, this trend changed between 1992 and 2013, with the Olympic finalists being remarkably older than in previous editions [26]. The dual career initiative in sport (which comprises the successful combination of education, training, or work with sport) can be an explanatory factor for swimmers peaking later in their careers [27]. This may also be dependent on the characteristics of each event (short-distance or long-distance). While for short-distance events there was a trend of an increase in the age of world record holders over the decades, an unchanged trend was found for long-distance ones. The consistent trend of swimmers achieving world records at older ages in short-distance events may be attributed to the requirements of the effort. These events typically rely heavily on explosive speed and power, which can be maintained or even improved with age through targeted training and experience [28]. The continued development of race strategy details [29], along with the improvement in novel recovery approaches or medical support [30], help older swimmers to extend their competitive careers and continue setting records at older ages. In contrast, the variations observed in the age of world record holders in long-distance events may reflect the unique physiological demands of these races. Long-distance events require a combination of endurance, aerobic capacity, and race-pacing maintenance [31], which may peak at an earlier stage of an athlete's career. Younger athletes may have a natural advantage in terms of aerobic capacity and recovery ability, allowing them to excel in longer-distance events more easily than their older counterparts [32].

Forecasting models of WRs (Figures 1 and 2) show promising evolution over the next decade (during the 2020–2029 period). As an example, predicted WRs for the 100 m freestyle event are 46.79 s and 46.69 s within the next few years. Three male WRs have already been broken within this time frame (e.g., David Popovici clocking 46.86 s in the 2022 European Aquatics Championships in Rome, Pan Zhanle clocking 46.80 s and 46.40 s in the 2024 World Aquatics Championships in Doha and in the Paris 2024 Olympics, respectively). This means that our models are fair and acceptable, and it should not be expected that swimming WRs will plateau so soon as previously reported by Nevill et al. [24]. There have already been 20 individual WRs (11 male and 9 female) in long-course events within this decade. Although there is a consensus that swimmers at this stage have already reached their physiological limits [33], there seems to be some room for improvement for further advancements in swimming performance. Considering the projections for the winning time in the Paris 2024 Olympics [34], it was not expected that any WR would be broken. Still, a fair number of WRs in freestyle events were achieved (two male WRs) and were somewhat in agreement with what our forecasts showed.

Forecasting models related to the age of record holders (Figures 3 and 4) showed us a mixed path. General stabilization in the age of world record holders in most of the freestyle events should be expected in the upcoming years. Stabilization in the age of male and female elite freestyle swimmers was seen between 1994 and 2016. Contrarily, Konig et al. [26] found an increase in the age of swimmers across World Championships and Olympic Games finals in a 20-year time frame (from 1992 to 2013). Although this might be a phenomenon with a random fluctuation, the age of a world record holder may sometimes not be within the age interval of those participating in the finals of great events. Some recent great performances were achieved by phenomenal teenage individuals like Summer McIntosh (15 years) and David Popovici (17 years). At the time of breaking the WRs, those swimmers were far from the 2016 average age (~23 years) of freestyle world-class swimmers [35]. So, this seems to be a random phenomenon with a nuanced pattern that opens space for outlier emergence that may counteract some forecast results. The existence of outliers is a well-known phenomenon that should be embraced by those who regularly work on sports prediction statistics [36]. As such, any future observations in this case should be interpreted with caution.

This study contributes valuable insights into the world records and age dynamics (e.g., career peak) for world-class swimmers. Coaches should be aware that there are sex-specific and event-specific dynamics in evolving trends over the decades. Understanding these trends can have implications for talent identification and long-term development of performance. This can also drive political strategies to keep the most talented swimmers close to the sport so that they can have a more sustainable career. The forecasting approach can also be seen as a powerful tool that can help us understand the potential performance and age limits for future world record breaking.

Some limitations and/or key aspects should be considered in upcoming research: (i) The included WRs were restricted to freestyle events. As such, a deeper analysis should be performed considering the progression of WRs in the remaining swimming events/techniques. (ii) Only WRs in the long-course pool were extracted, which leaves doubt as to whether the same trend would occur when WRs in the short-course pool are considered. (iii) Only broader trends about world records were shown, which suggests that future studies could delve into interdecadal analysis.

5. Conclusions

It can be concluded that a natural trend of improvement in WRs over the decades happened in long-course freestyle events, with world record holders becoming older mainly in short-distance events. The technology and training advancements played a key role here. In the upcoming years, WRs will improve, with distinct patterns being displayed in the age of world record holders depending on the competitive event. The age of male world record holders will stabilize in shorter events and decrease in longer distance ones, while for females, general stabilization should be expected for the majority of the competitive events.

Author Contributions: Conceptualization, M.J.C. and A.M.C.; methodology, M.J.C., L.Q.-N., S.F., A.M.C. and C.C.S.; formal analysis, M.J.C. and C.C.S.; data curation: L.Q.-N. and S.F.; writing—original draft preparation, M.J.C.; writing—review and editing, M.J.C., L.Q.-N., S.F., A.M.C. and C.C.S. All authors have read and agreed to the published version of the manuscript.

Funding: This work was supported by the Portuguese Foundation for Science and Technology (FCT), I.P., under the funding programs UIDP/05913/2020 (https://doi.org/10.54499/UIDB/05913/2020), UIDB04045/2020 and UIDP/MAT/00212/2020.

Institutional Review Board Statement: Not applicable.

Data Availability Statement: The data presented in this study are available on request from the corresponding author.

Conflicts of Interest: The authors declare no conflicts of interest.

References

1. Costa, M.J.; Marinho, D.A.; Santos, C.C.; Quinta-Nova, L.; Costa, A.M.; Silva, A.J.; Barbosa, T.M. The coaches' perceptions and experience implementing a Long-Term Athletic Development model in competitive swimming. *Front. Psychol.* **2021**, *12*, 685584. [CrossRef]
2. Lang, M.; Light, R. Interpreting and implementing the long term athlete development model: English swimming coaches' views on the (swimming) LTAD in practice. *Int. J. Sports Sci. Coach.* **2010**, *5*, 389–402. [CrossRef]
3. Costa, M.J.; Marinho, D.A.; Reis, V.M.; Silva, A.J.; Marques, M.C.; Bragada, J.A.; Barbosa, T.M. Tracking the performance of world-ranked swimmers. *J. Sports Sci. Med.* **2010**, *9*, 411–417.
4. Costa, M.J.; Marinho, D.A.; Bragada, J.A.; Silva, A.J.; Barbosa, T.M. Stability of elite freestyle performance from childhood to adulthood. *J. Sports Sci.* **2011**, *29*, 1183–1189. [CrossRef]
5. Brustio, P.R.; Cardinale, M.; Lupo, C.; Varalda, M.; De Pasquale, P.; Boccia, G. Being a top swimmer during the early career is not a prerequisite for success: A study on sprinter strokes. *J. Sci. Med. Sport.* **2021**, *24*, 1272–1277. [CrossRef]
6. Born, D.P.; Lomax, I.; Rüeger, E.; Romann, M. Normative data and percentile curves for long-term athlete development in swimming. *J. Sci. Med. Sport.* **2022**, *25*, 266–271. [CrossRef]
7. Allen, S.V.; Vandenbogaerde, T.J.; Hopkins, W.G. Career performance trajectories of Olympic swimmers: Benchmarks for talent development. *Eur. J. Sport. Sci.* **2014**, *14*, 643–651. [CrossRef]
8. Sokolovas, G. Biological maturation of swimmers. In *Proceedings of the VIIIth International Symposium on Biomechanics and Medicine in Swimming*; Keskinen, K.L., Komi, P.V., Hollander, A.P., Eds.; University of Jyväskylä: Jyväskylä, Finland, 1998; pp. 315–319.
9. Platonov, V. *Sport Training for High-Level Swimmers*; Phorte Editions: São Paulo, Brazil, 2005.
10. Hopkins, W.G.; Pike, J.C.; Nottle, C. Overall trends and individual trajectories of swimming performance in a decade of New Zealand National Championships. In *Proceedings of the XIth International Symposium on Biomechanics and Medicine in Swimming*; Norwegian School of Sport Science: Oslo, Norway, 2010; p. 72.
11. Baxter-Jones, A.D.; Sherar, L.B. Growth and maturation. In *Paediatric Exercise Physiology*; Armstrong, N., Ed.; Elsevier Limited: Amsterdam, The Netherlands, 2006; pp. 1–30.
12. Sandbakk, Ø.; Solli, G.S.; Holmberg, H.C. Sex differences in world-record performance: The influence of sport discipline and competition duration. *Int. J. Sports Physiol. Perform* **2018**, *13*, 2–8. [CrossRef]
13. Costa, A.M.; Breitenfeld, L.; Silva, A.J.; Pereira, A.; Izquierdo, M.; Marques, M.C. Genetic inheritance effects on endurance and muscle strength: An update. *Sports Med.* **2012**, *42*, 449–458. [CrossRef]
14. Chase, K.I.; Caine, D.J.; Goodwin, B.J.; Whitehead, J.R.; Romanick, M.A. A prospective study of injury affecting competitive collegiate swimmers. *Res. Sports Med.* **2013**, *21*, 111–123. [CrossRef]
15. Siekanska, M.; Blecharz, J. Transitions in the careers of competitive swimmers: To continue or finish with elite sport? *Int. J. Environ. Res. Public Health* **2020**, *17*, 6482. [CrossRef]

16. Knechtle, B.; Bragazzi, N.L.; König, S.; Nikolaidis, P.T.; Wild, S.; Rosemann, T.; Rüst, C.A. The age in swimming of champions in World Championships (1994–2013) and Olympic Games (1992–2012): A cross-sectional data analysis. *Sports* **2016**, *4*, 17. [CrossRef]
17. Rüst, C.A.; Knechtle, B.; Rosemann, T.; Lepers, R. The changes in age of peak swim speed for elite male and female Swiss freestyle swimmers between 1994 and 2012. *J. Sports Sci.* **2014**, *32*, 248–258. [CrossRef]
18. Ferguson, C.J. An effect size primer: A guide for clinicians and researchers. *Prof. Psychol. Res. Pract.* **2009**, *40*, 532–538. [CrossRef]
19. Lippi, G.; Banfi, G.; Favaloro, E.J.; Rittweger, J.; Maffulli, N. Updates on improvement of human athletic performance: Focus on world records in athletics. *Br. Med. Bull.* **2008**, *87*, 7–15. [CrossRef]
20. Barbosa, T.M.; Barbosa, A.C.; Simbaña Escobar, D.; Mullen, G.J.; Cossor, J.M.; Hodierne, R.; Arellano, R.; Mason, B.R. The role of the biomechanics analyst in swimming training and competition analysis. *Sports Biomech.* **2021**, *22*, 1734–1751. [CrossRef]
21. Yustres Amores, I.; Santos Del Cerro, J.; González-Mohíno, F.; Hermosilla, F.; González-Ravé, J.M. Modelling performance by continents in swimming. *Front. Physiol.* **2023**, *14*, 1075167. [CrossRef]
22. Colwin, C. Looking back, looking ahead. In *Breakthrough Swimming*; Champaign, I.L., Ed.; Human Kinetics: Champaign, IL, USA, 2002; pp. 217–228.
23. Biel, K.; Fischer, S.; Kibele, A. Kinematic analysis of take-off performance in elite swimmers: New OSB11 versus traditional starting blocks. In *XIth International Symposium for Biomechanics and Medicine in Swimming*; Norwegian School of Sport Science: Oslo, Norway, 2010; pp. 91–92.
24. Nevill, A.M.; Whyte, G.P.; Holder, R.L.; Peyrebrune, M. Are there limits to swimming world records? *Int. J. Sports Med.* **2007**, *28*, 1012–1017. [CrossRef]
25. Schulz, R.; Curnow, C. Peak performance and age among superathletes: Track and field, swimming, baseball, tennis, and golf. *J. Gerontol.* **1988**, *43*, 113–120. [CrossRef]
26. König, S.; Valeri, F.; Wild, S.; Rosemann, T.; Rüst, C.A.; Knechtle, B. Change of the age and performance of swimmers across World Championships and Olympic Games finals from 1992 to 2013: A cross-sectional data analysis. *SpringerPlus* **2014**, *3*, 652. [CrossRef]
27. Vidal-Vilaplana, A.; Valantine, I.; Staskeviciute-Butiene, I.; González-Serrano, M.H.; Capranica, L.; Calabuig, F. Combining sport and academic career: Exploring the current state of student-athletes' dual career research field. *J. Hosp. Leis. Sport Tour. Edu.* **2022**, *31*, 100399. [CrossRef]
28. Born, D.P.; Lorentzen, J.; Björklund, G.; Stöggl, T.; Romann, M. Variation vs. specialization: The dose-time-effect of technical and physiological variety in the development of elite swimmers. *BMC Res. Notes* **2024**, *17*, 48. [CrossRef]
29. Gonjo, T.; Olstad, B.H. Race Analysis in Competitive Swimming: A Narrative Review. *Int. J. Environ. Res. Public Health* **2020**, *18*, 69. [CrossRef]
30. Barry, L.; Lyons, M.; McCreesh, K.; Powell, C.; Comyns, T. The design and evaluation of an integrated training load and injury/illness surveillance system in competitive swimming. *Phys. Ther. Sport* **2023**, *60*, 54–62. [CrossRef]
31. López-Belmonte, Ó.; Ruiz-Navarro, J.J.; Gay, A.; Cuenca-Fernández, F.; Mujika, I.; Arellano, R. Analysis of pacing and kinematics in 3000 m freestyle in elite level swimmers. *Sport Biomech.* **2023**, 1–17, *advance online publication*. [CrossRef]
32. Mejias, J.E.; Bragada, J.A.; Costa, M.J.; Reis, V.M.; Garrido, N.D.; Barbosa, T.M. Young masters vs. elite swimmers: Comparison of performance, energetics, kinematics and efficiency. *Int. SportMed J.* **2014**, *15*, 165–177.
33. Costa, M.J.; Bragada, J.A.; Mejias, J.E.; Louro, H.; Marinho, D.A.; Silva, A.J.; Barbosa, T.M. Effects of swim training on energetics and performance. *Int. J. Sports Med.* **2013**, *34*, 507–513. [CrossRef]
34. Mujika, I.; Pyne, D.B.; Wu, P.P.; Ng, K.; Crowley, E.; Powell, C. Next-Generation models for predicting winning times in elite swimming events: Updated predictions for the Paris 2024 Olympic Games. *Int. J. Sports Physiol. Perfor.* **2023**, *18*, 1269–1274. [CrossRef]
35. Mallett, A.; Bellinger, B.; Derave, W.; Osborne, M.; Minahan, C. The age, height, and body mass of Olympic swimmers: A 50-year review and update. *Int. J. Sports Sci. Coach.* **2020**, *16*, 210–233. [CrossRef]
36. Aitken, T. Statistical analysis of top performers in sports with emphasis on the relevance of outliers. *Sports Eng.* **2004**, *7*, 75–88. [CrossRef]

applied sciences

MDPI

Article

Comparative Electromyographic Analysis in Leg Press of Traditional Fitness Equipment, Traditional Outdoor Fitness Equipment, and a New Model of Outdoor Fitness Equipment in Trained Young Men

Tomás Abelleira-Lamela [1], Pablo Jorge Marcos-Pardo [2,3,*], José Arturo Abraldes [4], Noelia González-Gálvez [1,*], Alejandro Espeso-García [1], Francisco Esparza-Ros [5] and Raquel Vaquero-Cristóbal [4]

1 Facultad del Deporte, UCAM Universidad Católica de Murcia, 30107 Murcia, Spain; tabelleira@ucam.edu (T.A.-L.); aespeso@ucam.edu (A.E.-G.)
2 SPORT Research Group (CTS-1024), CIBIS (Centro de Investigación para el Bienestar y la Inclusión Social), University of Almeria, 04120 Almeria, Spain
3 Department of Education, Faculty of Education Sciences, University of Almeria, 04120 Almeria, Spain
4 Research Group Movement Sciences and Sport (MS&SPORT), Department of Physical Activity and Sport Sciences, Faculty of Sport Sciences, University of Murcia, 30720 San Javier, Spain; abraldes@um.es (J.A.A.); raquel.vaquero@um.es (R.V.-C.)
5 Injury Prevention in Sport Research Group, Catholic University San Antonio of Murcia, 30107 Murcia, Spain; fesparza@ucam.edu
* Correspondence: pjmarcos@ual.es (P.J.M.-P.); ngonzalez@ucam.edu (N.G.-G.); Tel.: +34-950-01-53-52 (P.J.M.-P.); +34-968-27-88-24 (N.G.-G.)

Featured Application: The main practical application of this research is the presentation of a new type of outdoor fitness equipment (OFE) available to both trainers and users, with scientific support regarding its ability to recruit the desired muscles in a similar way to conventional gym machinery. In addition, it was possible to improve the effectiveness of the OFE by including an intensity regulation system, thus allowing the principle of intensity progression to be considered, as well as increasing the control of training intensity in this type of machinery.

Abstract: This study compares the electromyographic activity (EMG) of different muscle groups (rectus femoris, vastus lateralis, biceps femoris, tibialis anterior, and gastrocnemius) of the lower limbs when performing a traditional seated leg press (SLP) with a classic piece of outdoor fitness equipment (OFE-SLP), and with a new OFE leg press that allows the user to adjust the intensity of the exercise by means of a selectorized system (BIOFIT-LP). It was found that the EMG of the OFE-SLP was significantly lower than that of the SLP, but similar activations to those of the SLP were achieved with the BIOFIT-LP. In conclusion, the inclusion of a system to be able to change intensity of the exercise in OFE achieves an EMG activity similar to traditional machinery in trained young men.

Keywords: leg press; lower limb muscle; outdoor gym; resistance training; selectorized stack machine; surface EMG

1. Introduction

The practice of strength training has become popular in recent years, due to the various benefits it brings to the health and functionality of its practitioners [1–3]. Increased strength, hypertrophy, and metabolic improvements are some of the results sought through this type of training [1,4]. To carry this out, multiple resources can be employed, with the use of conventional gym equipment for strength training being one of the most common methods [4–6]. In addition, in recent years, a new line of outdoor fitness equipment (OFE) has emerged, which is sports equipment similar in operation to traditional gym equipment, but commonly located in outdoor public spaces such as parks [7,8]. It has become easier to find OFE due to its increasing popularity, especially among adults and older people who want to improve their health through outdoor strength training [8,9].

It is known that lower limb muscle mass and strength is a determining factor for multiple aspects such as functional capacity [10] and sports performance [11], with strength training of this part of the body being the quintessential means of skeletal muscle mass gain [12]. This has led to a proliferation of machinery to exercise the lower limb musculature, most of them being multi-joint [11,13,14] and involving different muscle groups [11].

However, prior to the use of any type of machinery for strength work, it is necessary to analyze whether it is effective, i.e., whether it achieves a sufficient activation to generate adaptations of the musculature involved, and whether it is safe, i.e., whether it is not harmful to the musculoskeletal and joint health of the exercise performer [15]. In terms of efficacy, techniques such as surface electromyographic analysis (sEMG) are used to determine the musculature involved in the exercise action. This tool allows the study of muscle function through electrophysiological recording techniques that detect the electrical potentials that cross the membrane of the fibers in the superficial muscles involved in each exercise [16]. In this way, electromyographic activity (EMG) makes it possible to discover the degree of neuromuscular activation during the execution of an exercise for each of the muscle groups that are analyzed, demonstrating its efficacy [17,18].

Many studies have analyzed the EMG of the different muscle groups of the lower limbs in exercises performed with conventional gym equipment [19,20], finding that guided machines such as the traditional seated leg press (SLP) are effective for the work of this muscle group, with activation especially observed in the quadriceps femoris muscles [21]. In this type of equipment, it has also been shown that submitting the subject to a greater intensity generates a greater EMG of the involved muscles [22]. However, despite the popularity of OFE and, in many cases, the large investment of public funds being made for the purchase and installation of these machines in public spaces [8,23], thus far it has only been found that their use can improve both the body composition and strength production capacity in legs and arms in the older population [9], and no studies have analyzed whether OFE is effective, i.e., whether it involves the target musculature at an intensity sufficient to generate structural adaptations in the muscle. On the other hand, classical OFE does not allow progression of the intensity of the exercise, as it allows work exclusively with a self-body mass as the unique intensity of the exercise [24]. Therefore, it would be interesting to also analyze the EMG with a new line of OFE that does allow an increase in intensity of the exercise, with a system similar to that of traditional gym equipment. However, so far there are no data available in this regard either.

Therefore, the objectives of this research were: (a) to analyze and compare the EMG of the lower limb muscle groups when performing an SLP with traditional gym equipment, a leg press machine of classic OFE (OFE-SLP), and a leg press machine belonging to a new line of OFE which allows an adjustment in the intensity of the exercise by means of a selectorized system (BIOFIT-LP); and (b) to determine the differences in the EMG of the lower limb muscle groups at 60 and 75% of the 1RM in the machines that allow the selection of the intensity of the exercise (SLP and BIOFIT-LP).

Regarding the hypotheses of the present investigation, it is hypothesized that: (a) there will be an EMG activation of the lower limb muscles when performing the SLP in the three machines, with the activation being greater when using the machines that allow increasing the intensity of the exercise (SLP and BIOFIT-LP) than in those that only allow working with a self-body weight as the unique intensity of the exercise (OFE-SLP). This could imply that the SLP and BIOFIT-LP machines are more effective for lower limb strength training than the OFE-SLP machine. It is also hypothesized that the activation will be similar between the machines that allow selecting the intensity of the exercise (SLP vs. BIOFIT-LP) for the major muscle groups involved, with differences in some of the synergists. This would imply that both could be equally effective in working on lower limb strength in general, with the differences in EMG activation found between both machines being due to biomechanical differences in their execution. Because of the above, the machinery to work with could be selected according to the synergists one prefers to involve to a greater extent; and (b) the greater the intensity of the exercise on the machine, the greater the EMG of the muscles involved in both machines. The consequence of this hypothesis, if it is proven, would be that the intensity of the exercise can be used as another variable in the estimation of the training load, regardless of the model, in the machines that allow a selection of the load.

2. Materials and Methods

2.1. Experimental Approach to the Problem

A randomized cross-sectional study was carried out. It consisted of a cross-sectional study where the 1RM was calculated on the seated leg press (SLP) (Figure 1a) and the new BIOFIT leg press (BIOFIT-LP) (Figure 1b) (session 1 and session 2, 72 h apart), and the EMG activity of five muscle groups was analyzed on three different types of leg press machines (session 3, 72 h after session 2). With respect to the machines chosen, one belongs to a line commonly used in the fitness field, more specifically the SLP model (Technogym; Cesena, Italy), which allows adjusting the intensity of the exercise through the use of a selectorized system [4] (Figure 1a); another is from a classic OFE machine line, corresponding with the same machine model, the Columpio (OFE-SLP) (Entorno Urbano, Murcia, Spain) (Figure 1c), which only allows exercising with self-weight as the unique intensity of the exercise (Chow et al., 2021) [25]; and a third one belongs to a new line of OFE, corresponding to the same machine model, called BIOFIT-LP (Entorno Urbano, Murcia, Spain), which allows adjustment of the intensity of the exercise by means of a selectorized system [4], and which is currently under patent registration (patent registration for the OFE-SLP at the Spanish Patent and Trademark Office, application number: ES2975897; and patent registration for the selector of the intensity of the exercise, Spanish Patent and Trademark Office, application number: ES2975886) (Figure 1b). The operation of this machine is based on two handles to be taken on both sides of the body, starting the movement from a 90° flexion of the knee, and from here one must complete a full knee extension against resistance and against gravity (concentric phase). The intensity of the exercise is selected with a selectorized system. On the other hand, the eccentric phase of this exercise goes from knee extension to 90° knee flexion, a descent movement that is done in favor of gravity (eccentric phase). The study utilized a specific research model, as illustrated in Supplementary Table S1.

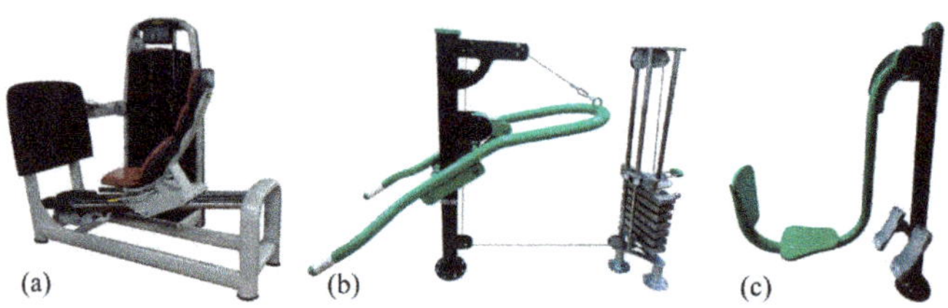

Figure 1. Fitness equipment used. (**a**) Seated leg press (Technogym; Cesena, Italia); (**b**) outdoor fitness equipment leg press with selectorized system (Entorno Urbano, Murcia, Spain); (**c**) outdoor fitness equipment seated leg press (Entorno Urbano, Murcia, Spain).

The study was designed according to the principles of the Declaration of Helsinki and with the prior approval of the institutional ethics committee of the Universidad Católica de Murcia (Ethical Application Ref: CE111908). All participants were informed of the tests to be performed and voluntarily signed a consent form prior to the start of the study.

2.2. Subjects

To calculate the sample size, we used the methodology from previous studies based on the standard deviation (SD) [26]. The Rstudio 3.15.0 software was used for this purpose. Sample size and power were established according to the standard deviation of the peak

EMG activity of the biceps femoris variable (SD = 49.68 μV) presented by a sample of similar characteristics, composed of young males (n = 10) [27]. With an estimated error (d) of 30.79 μV, a confidence interval of 95% (95% CI), a power of 95%, and a significance level of 0.05, the minimum sample needed for conducting the study was 34 subjects.

The sample consisted of 34 males (mean age = 21.54 ± 1.80 years; mean body mass = 74.27 ± 9.99 kg; mean height = 176.62 ± 6.67 cm). The inclusion criteria were: (1) having at least 2 years of experience in training with gym machines and currently training with this methodology at least three times a week; (2) being between 19 and 24 years old; and (3) being male. The exclusion criteria were: (1) having exercised in the 48 h prior to the first session or between evaluation sessions [20]; (2) having suffered any lower extremity, hip, or back injury in the last 6 months; and (3) having any known muscle abnormality or joint disease.

2.3. *Measurements*

2.3.1. Questionnaire

All participants completed an ad hoc questionnaire to obtain their sociodemographic characteristics, as well as their experience in sports and in the field of machine training and the injuries and pathologies suffered.

2.3.2. Anthropometric Characteristics

Anthropometric measurements of body mass and height were taken following the indications of the International Society for the Advancement of Kinanthropometry (ISAK) [28] by an ISAK-accredited level 2 anthropometrist. A Tanita BC-545N scale (Tanita, Arlington Heights, IL, USA) and an HR001 portable stadiometer (Tanita, Arlington Heights, IL, USA) were used.

2.3.3. Determination of 1RM

The calculation of the 1RM was performed on the SLP and BIOFIT-LP machine, both of which have a selectorized system that allows the intensity of the exercise to be modified. This test was not performed with the OFE-SLP as it does not have any system for increasing the external intensity of the exercise. For the 1RM tests, the order of execution of the SLP and BIOFIT-LP was randomized. In both cases, the protocol proposed by the National Strength and Conditioning Association [29] was employed. Volunteers warmed up by performing a set with an intensity of the exercise with which they could perform between 5 and 10 repetitions. After 1 min of rest, the intensity of the exercise was increased by 10 to 20%, an intensity of the exercise with which the participant had to perform a minimum of 3 and a maximum of 5 repetitions. After 2 min of rest, the intensity of the exercise was increased once again by 10 to 20%. With this intensity of the exercise, the participant had to perform a minimum of 2 and a maximum of 3 repetitions. After this, the intensity of the exercise was increased again by 10 to 20% to perform one repetition, with this weight increased or reduced by 5 and 10%, depending on whether it was lifted correctly or not, until the intensity of the exercise corresponding to the 1RM was obtained. This required a maximum of 5 attempts with different intensity of the exercises, with 4 min of rest between each of them [29]. After 72 h, the same protocol was performed on the other machine, according to the previous randomization performed.

2.3.4. Electromyographic (EMG) Analysis

In a third measurement session, separated by 72 h from the previous session, the EMG of the quadriceps femoris (rectus femoris and vastus lateralis), biceps femoris (long head), tibialis anterior, and gastrocnemius (medial head) muscle groups was measured when performing 5 repetitions on the SLP and BIOFIT-LP machines at 60 and 75% of 1RM. These intensities were chosen as they are commonly used in different training programs, as well as in different EMG studies with guided machinery [30]. The OFE-SLP EMG was analyzed with a single intensity of the exercise allowed (self-weight).

For electrode placement, the different skin areas where the electrodes would be placed were prepared by shaving the hair and cleaning the skin with 96% alcohol and sterile gauze to reduce skin impedance [11,18]. Gelled disposable surface electrodes were used, placed with a 2 cm center-to-center distance, aligned with the longitudinal axis of the muscle fibers, following the methodology from previous studies [31]. The electrodes were placed according to the recommendations of the Surface Electromyography for the Non-invasive Assessment of Muscles (SENIAM) [31] for each of the muscles measured: quadriceps femoris (rectus femoris and vastus lateralis), biceps femoris (long head), tibialis anterior, and gastrocnemius (medial head) (Figure 2). Consistent with similar research, bilateral symmetry was assumed during exercise performance, so electrode placement on the right side of the body was standardized [32]. EMG activity during the exercises was recorded using a surface electromyography system at a sampling rate of 1500 Hz (MuscleLab Ergotest Innovation AS, Stathelle, Norway). The electromyographic signal during the calculation of the 1RM was not taken, since the variability that the EMG signal would present due to fatigue because of the procedure used following the indications of the National Strength and Conditioning Association could lead to bias. In addition, as the sessions in which 1RM was calculated and the sessions in which EMG data were collected were different, variables such as skin impedance or hydration could include more intra-subject variability [11,18]. Therefore, it was decided to measure EMG activation only on the third day, which was the day on which the performance on the different machines was compared.

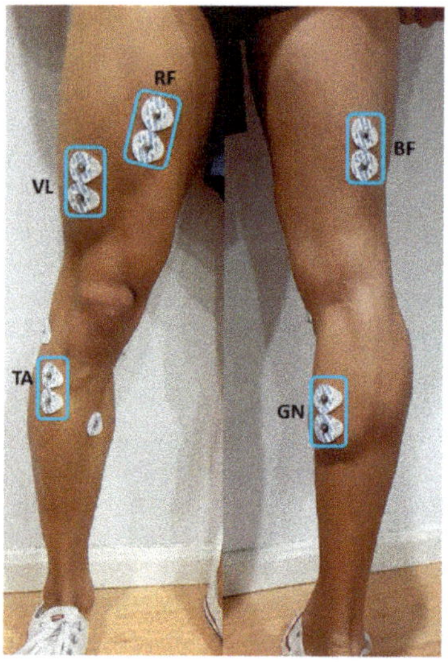

Figure 2. Electrode location on the right leg (**left**—anterior view; **right**—posterior view). RF—rectus femoris, VL—vastus lateralis, BF—biceps femoris, TA—tibialis anterior, GN—gastrocnemius.

Prior to the study, intra-instrument reliability was assessed with a group of 10 training young men by means of test–retest. The tests described above were performed two times in the same session on the same volunteers [33]. It was determined that the electromyographic analysis used in the present study was highly reproducible with a coefficient of variation between 7 and 14% for all muscles on all machines, confirming that the measurement used was sufficiently reproducible.

2.3.5. Determination of Concentric and Eccentric Phases

To differentiate the concentric and eccentric phases during exercise execution, a MuscleLab twin axial electrogoniometer was used (Ergotest Innovation AS, Stathelle, Norway). It was placed on the subject's left leg at the knee joint on the lateral outer side. The distal gauge of the goniometer was aligned with respect to the leg, while the fixed gauge was aligned with respect to the thigh [34]. It was calibrated using a hand-held goniometer at 90° of knee flexion in stable relaxed sedentation. The point of maximum knee extension was checked in situ, thus differentiating the concentric phase as those data recorded from the beginning at 90° of knee flexion up to that maximum knee extension, and the eccentric phase as the data recorded from maximum knee extension to 90° of knee flexion. The complete phase consisted of the analysis of the entire joint range recorded from 90° of onset to 90° of end [35].

2.4. Randomization and Blinding

Following the methodology from previous studies, the order of execution of the machines for 1RM analysis, of execution of the machines for EMG analysis, and of the 1RM intensity at which each machine was to be started were randomized [27]. This process was performed by the principal investigator in the presence of investigators external to the present research, using a computer-generated random number table.

All measurements were performed after randomization. The investigators who performed the EMGs did not participate in the 1RM calculation tests. In the measurements, the two investigators involved in the data collection were also cross-blinded, whereby one controlled the exercise execution technique and provided the guidelines for the development of the protocol, and the other investigator was only in charge of verifying the correct recording of the EMG data.

2.5. Procedure

Figure 3 shows in schematic form the procedures carried out during the three sessions. In session 1, the volunteers self-completed the ad hoc questionnaire, followed by anthropometric measurements. After this, the key points for the correct execution on each of the three machines (SLP, OFE-SLP, and BIOFIT-LP) were explained to them, in order to familiarize the participants with the expected execution technique on each of them for data collection. More specifically, an indication was given to them that the executions on the different machines should include the full range of motion from a 90° knee flexion to full extension, keeping the sole of the foot fixed in position and maintaining contact with the floor/platform for the duration of the repetition.

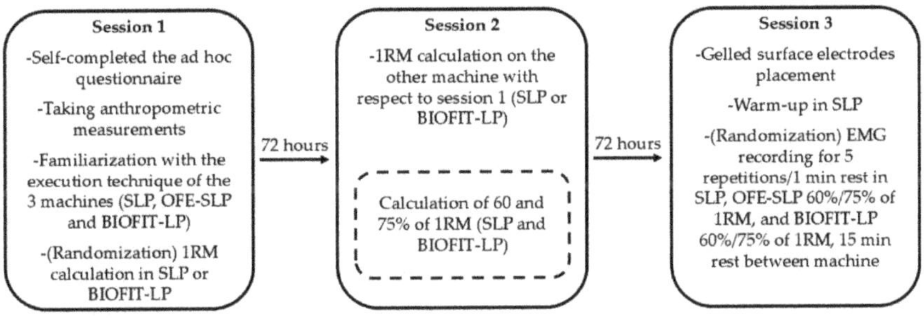

Figure 3. Flow diagram.

After this, the subjects performed the procedure for the calculation of the 1RM on the SLP or BIOFIT-LP machines, whose execution order was established randomly. In session 2, the process was repeated for the other machine. With the data obtained with this procedure, the 60 and 75% values of the 1RM intensity of the exercise were subsequently calculated for each subject on both machines.

In session 3, after electrode placement, participants performed a warm-up set consisting of 12 repetitions at 30% of 1RM, 10 repetitions at 50% of 1RM, 8 repetitions at 70% of 1RM, and 2 repetitions at 90% of 1 RM on the SLP machine, as the subjects had the most previous experience with it, as in similar studies [36]. Subsequently, the EMG measurement was started in the machine assigned randomly (SLP, OFE-SLP, or BIOFIT-LP). In the case of SLP and BIOFIT-LP machines, the intensity was also randomized (60% or 75% of 1RM). The participant was asked to perform 5 repetitions on the first machine and at the first intensity percentage, with a constant speed throughout [11]. A rest of 1 min was given between each of the repetitions to avoid the effect of fatigue. After 15 min, the same protocol was performed on the same machine, but with the other intensity percentage. After finishing the entire protocol with the first machine, a 15 min rest was provided, after which the protocol was started with the second machine, following the same procedure described above [11]. In the OFE-SLP case, 1 set of 5 repetitions was performed following the procedure previously described with the participant's self-weight, as the participant's intensity of the exercise could not be adjusted [25]. Resting times were set following previous methodologies, so that fatigue would not affect the results [22]. For the analysis of the data, the methodology of a previous study was followed [11]; the first and last repetitions were discarded, and the repetition with the highest root mean square (RMS) of the main muscle group of the movement, the rectus femoris, was selected, provided that it had been performed correctly, with a recording performed in each of the muscles involved [14,37,38].

All measurement sessions were separated by 72 h to avoid possible fatigue effects [37]. All subjects were summoned to all sessions at 8:00 a.m. The measurements were performed in the same room, with a standardized temperature of 20 °C and constant humidity conditions (60%).

2.6. Statistical Analysis

All data were analyzed using the Musclelab v10.5.67 software (MuscleLab Ergotest Innovation AS, Stathelle, Norway) by passing the raw amplified EMG signal to RMS [14,38]. Thus, the RMS of the peak EMG activity variable was taken in the complete repetition, and in its concentric and eccentric phases, for each muscle group, for the subsequent comparison between machines, as in previous studies [39]. The statistical analysis of the data was performed using SPSS software (Version 25; IBM Corporation, Armonk, NY, USA). The Kolmogorov–Smirnov test and Mauchly's W test were used to evaluate the normality of the data. The mean and standard deviation were calculated. A repeated-measures ANOVA and Bonferroni's correction were used to compare the different muscle activations between machines. To protect against a Type I error, Bonferroni's correction was used to achieve $p = 0.005$ for statistical significance. The effect size was calculated using partial eta-squared (η^2p) for analysis of variance and will be defined as small: ES $\geq$ 0.10; moderate: ES $\geq$ 0.30; large: ES $\geq$ 1.2; or very large: ES $\geq$ 2.0. An error of $p < 0.05$ was established.

3. Results

Table 1 shows the 1RM strength test results for the traditional SLP and BIOFIT-LP machines.

Table 1. Results of the 1RM strength test.

	1RM	60% 1RM	75% 1RM
SLP (kg)	201.51 ± 35.35	120.91 ± 21.21	151.14 ± 26.51
BIOFIT-LP (kg)	56.31 ± 13.72	31.85 ± 11.27	39.82 ± 14.09

1RM = one-repetition maximum; BIOFIT-LP = BIOFIT leg press; SLP = traditional seated leg press.

Table 2 and Figure 4 show the mean and standard deviation of the root mean square recorded in each of the muscles analyzed during the use of the SLP at 60 and 75% of 1RM, the BIOFIT-LP at 60 and 75% of 1RM, and the OFE-SLP with self-weight as intensity of the exercise. In all cases, it was found that the vastus lateralis was the muscle group that attained the greatest activation, followed by the rectus femoris, and thirdly by the tibialis anterior. After these, in the case of SLP and BIOFIT-LP, the next muscle groups with the highest activation were the biceps femoris and gastocnemius, in that order; while with the OFE-SLP machine, the order of the latter was reversed.

Table 2. Root mean square (µV) recording of the electromyographic signal in the different machines with different 1RM intensities.

	VL (M ± SD)	RF (M ± SD)	BF (M ± SD)	TA (M ± SD)	GN (M ± SD)
Complete					
SLP 60%	199.17 ± 61.46	134.96 ± 50.57	64.77 ± 38.23	68.29 ± 37.92	61.43 ± 32.66
SLP 75%	241.01 ± 70.83	176.98 ± 72.18	82.17 ± 39.51	88.08 ± 37.27	77.65 ± 37.31
BIOFIT-LP 60%	219.09 ± 69.23	149.25 ± 64.14	57.51 ± 33.48	133.69 ± 63.86	39.48 ± 20.78
BIOFIT-LP 75%	235.99 ± 75.52	162.60 ± 67.85	67.91 ± 36.25	140.83 ± 80.50	43.11 ± 19.41
OFE-SLP	66.89 ± 26.28	30.00 ± 17.76	21.22 ± 10.04	58.18 ± 55.18	40.18 ± 26.93
Concentric					
SLP 60%	224.98 ± 66.44	149.84 ± 60.63	77.11 ± 38.01	81.63 ± 46.72	75.83 ± 41.86
SLP 75%	274.71 ± 80.92	206.55 ± 88.28	95.83 ± 43.83	100.44 ± 49.77	90.85 ± 40.21
BIOFIT-LP 60%	251.12 ± 80.71	157.21 ± 74.59	69.14 ± 30.52	104.11 ± 51.47	48.86 ± 26.96
BIOFIT-LP 75%	271.15 ± 91.27	175.80 ± 74.10	82.08 ± 38.36	120.86 ± 67.79	50.45 ± 22.55
OFE-SLP	75.26 ± 32.99	28.09 ± 13.92	25.92 ± 12.39	38.52 ± 31.77	49.11 ± 32.26
Eccentric					
SLP 60%	168.86 ± 66.04	114.56 ± 56.17	39.28 ± 26.53	49.32 ± 31.60	41.51 ± 23.71
SLP 75%	196.34 ± 70.17	138.85 ± 60.05	50.26 ± 27.46	68.47 ± 30.36	53.89 ± 27.49
BIOFIT-LP 60%	179.06 ± 59.44	138.64 ± 63.56	26.94 ± 11.79	151.36 ± 89.07	24.41 ± 13.89
BIOFIT-LP 75%	187.59 ± 60.51	146.03 ± 68.19	32.87 ± 15.79	153.58 ± 104.76	30.84 ± 18.44
OFE-SLP	57.08 ± 27.61	28.70 ± 19.35	16.00 ± 9.36	64.52 ± 70.84	30.65 ± 23.68

BF = biceps femoris; BIOFIT-LP = BIOFIT leg press; GN = gastrocnemius; OFE-SLP = classic OFE; RF = rectus femoris; SLP = traditional seated leg press; TA = tibialis anterior; VL = vastus lateralis.

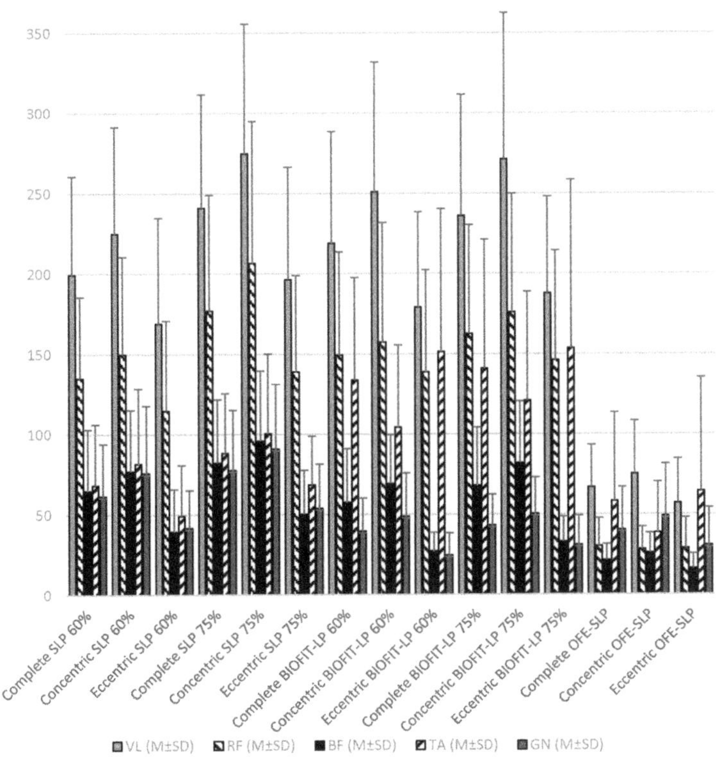

Figure 4. Root mean square (µV) representation of the electromyographic signal in the different machines with different 1RM intensities.

Table 3 shows the differences between machines and between intensity percentages in the complete run, Table 4 shows the concentric phase, and Table 5 shows the eccentric phase. No significant differences were found in the activation of any muscle group when comparing the SLP and BIOFIT-LP machines at the same intensity (60% 1RM vs. 60% 1RM or 75% 1RM vs. 75% 1RM), neither in the complete run, in the concentric phase, nor in the eccentric phase ($p > 0.05$), except for the biceps femoris in the eccentric phase at 75% 1RM, which showed a greater significant activation in SLP as compared to BIOFIT-LP (ES = 0.60; moderate effect size; $p < 0.01$); in the tibialis anterior, where there was a greater activation at both intensities in the BIOFIT-LP machine in the full stroke (ES = 0.53–0.86; moderate effect size; $p < 0.01$–0.001) and eccentric phase (ES = 0.69–0.89; moderate effect size; $p < 0.001$), but not in the concentric phase (ES = 0.3–0.5; moderate effect size; $p > 0.05$); and gastrocnemius, which showed a greater activation in SLP than in BIOFIT-LP at both intensities and all phases (ES = 0.68–0.91; moderate effect size; $p < 0.05$–0.001).

Table 3. Differentiation of peak electromyographic activity (μV) between different machines and intensities of 1RM during a full execution.

Complete Repetition		SLP 60%	SLP 75%	BIOFIT-LP 60%	BIOFIT-LP 75%	OFE-SLP
Vastus Lateralis	SLP 60%	-	−41.84 ± 4.48 †	−19.92 ± 8.38	−36.82 ± 9.80 **	132.285 ± 8.82 †
	SLP 75%	41.84 ± 4.48 †	-	21.91 ± 8.16	5.01 ± 9.93	174.12 ± 10.91 †
	BIOFIT-LP 60%	19.92 ± 8.38	−21.91 ± 8.16	-	−16.90 ± 5.03 *	152.21 ± 9.76 †
	BIOFIT-LP 75%	36.82 ± 9.80 **	−5.01 ± 9.93	16.90 ± 5.03 *	-	169.11 ± 10.82 †
	OFE-SLP	−132.29 ± 8.82 †	−174.12 ± 10.91 †	−152.21 ± 9.76 †	−169.11 ± 10.82 †	-
Rectus Femoris	SLP 60%	-	−42.02 ± 5.84 †	−14.30 ± 8.89	−27.64 ± 10.59	104.96 ± 8.03 †
	SLP 75%	42.02 ± 5.84 †	-	27.73 ± 10.74	14.38 ± 13.28	146.98 ± 11.84 †
	BIOFIT-LP 60%	14.30 ± 8.89	−27.73 ± 10.74	-	−13.34 ± 6.40	119.26 ± 10.04 †
	BIOFIT-LP 75%	27.64 ± 10.59	−14.38 ± 13.28	13.34 ± 6.40	-	132.60 ± 10.86 †
	OFE-SLP	−104.96 ± 8.03 †	−146.98 ± 11.84 †	−119.26 ± 10.04 †	−132.60 ± 10.86 †	-
Biceps Femoris	SLP 60%	-	−17.40 ± 3.41 †	7.25 ± 5.70	−3.14 ± 6.75	43.55 ± 6.82 †
	SLP 75%	17.40 ± 3.41 †	-	24.65 ± 5.99 **	14.26 ± 6.80	60.94 ± 6.75 †
	BIOFIT-LP 60%	−7.25 ± 5.70	−24.65 ± 5.99 **	-	−10.39 ± 2.83 *	36.29 ± 6.11 †
	BIOFIT-LP 75%	3.14 ± 6.75	−14.26 ± 6.80	10.39 ± 2.83 *	-	46.68 ± 6.44 †
	OFE-SLP	−43.55 ± 6.82 †	−60.94 ± 6.75 †	−36.29 ± 6.11 †	−46.68 ± 6.44 †	-
Tibialis Anterior	SLP 60%	-	−19.78 ± 3.35 †	−65.39 ± 10.69 †	−72.54 ± 13.67 †	10.12 ± 9.63
	SLP 75%	19.78 ± 3.35 †	-	−45.61 ± 10.58 **	−52.75 ± 12.87 **	29.90 ± 10.28
	BIOFIT-LP 60%	65.39 ± 10.69 †	45.61 ± 10.58 **	-	−7.14 ± 8.37	75.51 ± 10.48 †
	BIOFIT-LP 75%	72.54 ± 13.67 †	52.75 ± 12.87 **	7.14 ± 8.37	-	82.65 ± 14.59 †
	OFE-SLP	−10.12 ± 9.63	−29.90 ± 10.28	−75.51 ± 10.48 †	−82.65 ± 14.59 †	-
Gastrocn-emius	SLP 60%	-	−16.22 ± 3.80 **	21.96 ± 5.64 **	18.32 ± 5.72 *	21.25 ± 5.78 **
	SLP 75%	16.22 ± 3.80 **	-	38.17 ± 5.93 †	34.54 ± 6.42 †	37.46 ± 6.87 †
	BIOFIT-LP 60%	−21.96 ± 5.64 **	−38.17 ± 5.93 †	-	−6.63 ± 2.40	−0.71 ± 5.56
	BIOFIT-LP 75%	−18.32 ± 5.72 *	−34.54 ± 6.42 †	6.63 ± 2.40	-	2.93 ± 5.28
	OFE-SLP	−21.25 ± 5.78 **	−37.46 ± 6.87 †	0.71 ± 5.56	−2.93 ± 5.28	-

BIOFIT-LP = BIOFIT leg press; OFE-SLP = classic OFE; SLP = seated leg press; * = p value < 0.05; ** = p value < 0.01; † = p value < 0.001.

Table 4. Differentiation of peak electromyographic activity (μV) between different machines and 1RM intensities during the concentric phase.

	Concentric Phase	SLP 60%	SLP 75%	BIOFIT-LP 60%	BIOFIT-LP 75%	OFE-SLP
Vastus Lateralis	SLP 60%	-	−49.73 ± 7.90 †	−26.14 ± 11.69	−46.17 ± 14.00 *	149.72 ± 10.35 †
	SLP 75%	49.73 ± 7.90 †	-	23.59 ± 10.00	3.56 ± 13.40	199.45 ± 13.69 †
	BIOFIT-LP 60%	26.14 ± 11.69	−23.59 ± 10.00	-	−20.03 ± 6.41 *	175.86 ± 13.00 †
	BIOFIT-LP 75%	46.17 ± 14.00 *	−3.56 ± 13.40	20.03 ± 6.41 *	-	195.89 ± 14.67 †
	OFE-SLP	−149.72 ± 10.35 †	−199.45 ± 13.69 †	−175.86 ± 13.00 †	−195.89 ± 14.67 †	-
Rectus Femoris	SLP 60%	-	−56.72 ± 8.37 †	−7.37 ± 11.84	−25.96 ± 14.09	121.75 ± 10.01 †
	SLP 75%	56.72 ± 8.37 †	-	49.35 ± 14.86 *	30.76 ± 17.70	178.46 ± 15.09 †
	BIOFIT-LP 60%	7.37 ± 11.84	−49.35 ± 14.86 *	-	−18.59 ± 8.61	129.11 ± 12.32 †
	BIOFIT-LP 75%	25.96 ± 14.09	−30.76 ± 17.70	18.59 ± 8.61	-	147.71 ± 12.19 †
	OFE-SLP	−121.75 ± 10.01 †	−178.46 ± 15.09 †	−129.11 ± 12.32 †	−147.71 ± 12.19 †	-
Biceps Femoris	SLP 60%	-	−18.72 ± 4.68 **	7.97 ± 7.31	−4.98 ± 8.76	51.19 ± 7.05 †
	SLP 75%	18.72 ± 4.68 **	-	26.69 ± 7.99 *	13.75 ± 8.90	69.91 ± 7.75 †
	BIOFIT-LP 60%	−7.97 ± 7.31	−26.69 ± 7.99 *	-	−12.94 ± 3.82 *	43.22 ± 5.72 †
	BIOFIT-LP 75%	4.98 ± 8.76	−13.75 ± 8.90	12.94 ± 3.82 *	-	56.17 ± 7.00 †
	OFE-SLP	−51.19 ± 7.05 †	−69.91 ± 7.75 †	−43.22 ± 5.72 †	−56.17 ± 7.00 †	-
Tibialis Anterior	SLP 60%	-	−18.82 ± 3.91 †	−22.49 ± 9.25	−39.24 ± 13.60	43.11 ± 8.65 †
	SLP 75%	18.82 ± 3.91 †	-	−3.67 ± 8.61	−20.42 ± 12.53	61.92 ± 9.58 †
	BIOFIT-LP 60%	22.49 ± 9.25	3.67 ± 8.61	-	−16.75 ± 9.42	68.59 ± 9.80 †
	BIOFIT-LP 75%	39.24 ± 13.60	20.42 ± 12.53	16.75 ± 9.42	-	82.34 ± 13.03 †
	OFE-SLP	−43.11 ± 8.65 †	−61.92 ± 9.58 †	−68.59 ± 9.80 †	−82.34 ± 13.03 †	-
Gastrocn-emius	SLP 60%	-	−15.02 ± 4.80 *	26.96 ± 7.39 *	25.38 ± 7.21 *	26.72 ± 7.11 **
	SLP 75%	15.02 ± 4.80 *	-	41.98 ± 6.56 †	40.40 ± 6.91 †	41.74 ± 6.85 †
	BIOFIT-LP 60%	−26.96 ± 7.39 *	−41.98 ± 6.56 †	-	−1.59 ± 3.29	−0.24 ± 6.95
	BIOFIT-LP 75%	−25.38 ± 7.21 *	−40.40 ± 6.91 †	1.59 ± 3.29	-	1.35 ± 6.40
	OFE-SLP	−26.72 ± 7.11 **	−41.74 ± 6.85 †	0.24 ± 6.95	−1.35 ± 6.40	-

BIOFIT-LP = BIOFIT leg press; OFE-SLP = classic OFE; SLP = seated leg press; * = p value < 0.05; ** = p value < 0.01; † = p value < 0.001.

When comparing the OFE-SLP with SLP and the OFE-SLP with BIOFIT-LP, a significantly lower activation was found in the OFE-SLP for all muscle groups, in the full stroke, concentric, and eccentric phases (ES = 0.69–1.08; moderate effect size; $p < 0.01$–0.001), except for the tibialis anterior as compared to SLP for the full stroke and eccentric phase (ES = 0.09–0.42; effect size between small and moderate; $p > 0.05$); for the gastrocnemius as compared to BIOFIT-LP for the full stroke, concentric phase, and eccentric phase (ES = 0.09–0.2; small effect size; $p > 0.05$); and for the gastrocnemius as compared to SLP at 60% 1RM in the eccentric phase (ES = 0.25; small effect size; $p > 0.05$).

When comparing the OFE-SLP with SLP and the OFE-SLP with BIOFIT-LP, a significantly lower activation was found in the OFE-SLP for all muscle groups, in the full stroke, concentric, and eccentric phases (ES = 0.2–0.5 effect size between small and moderate; $p < 0.01$–0.001), except for the tibialis anterior, as compared to SLP for the full stroke and eccentric phase (ES = 0.14–0.5; effect size between small and moderate; $p > 0.05$); for the gastrocnemius, as compared to BIOFIT-LP for the full stroke, concentric phase, and eccentric phase (ES = 0.2–0.9; effect size between small and moderate; $p > 0.05$); and for the gastrocnemius as compared to SLP at 60% 1RM in the eccentric phase (ES = 0.25; small effect size; $p > 0.05$).

Table 5. Differentiation of peak electromyographic activity (μV) between different machines and 1RM intensities during the eccentric phase.

Eccentric Phase		SLP 60%	SLP 75%	BIOFIT-LP 60%	BIOFIT-LP 75%	OFE-SLP
Vastus Lateralis	SLP 60%	-	−27.48 ± 3.49 †	−10.20 ± 7.63	−18.73 ± 7.11	111.79 ± 8.78 †
	SLP 75%	27.48 ± 3.49 †	-	17.28 ± 8.74	8.75 ± 8.02	139.27 ± 9.61 †
	BIOFIT-LP 60%	10.20 ± 7.63	−17.28 ± 8.74	-	−8.53 ± 4.09	121.99 ± 7.41 †
	BIOFIT-LP 75%	18.73 ± 7.11	−8.75 ± 8.02	8.53 ± 4.09	-	130.52 ± 7.55 †
	OFE-SLP	−111.79 ± 8.78 †	−139.27 ± 9.61 †	−121.99 ± 7.41 †	−130.52 ± 7.55 †	-
Rectus Femoris	SLP 60%	-	−24.29 ± 6.01 **	−24.08 ± 11.08	−31.47 ± 11.64	85.86 ± 8.64 †
	SLP 75%	24.29 ± 6.01 **	-	0.21 ± 10.15	−7.18 ± 11.48	110.15 ± 9.63 †
	BIOFIT-LP 60%	24.08 ± 11.08	−0.21 ± 10.15	-	−7.39 ± 6.28	109.94 ± 9.10 †
	BIOFIT-LP 75%	31.47 ± 11.64	7.18 ± 11.48	7.39 ± 6.28	-	117.33 ± 10.15 †
	OFE-SLP	−85.86 ± 8.64 †	−110.15 ± 9.63 †	−109.94 ± 9.10 †	−117.33 ± 10.15 †	-
Biceps Femoris	SLP 60%	-	−10.97 ± 2.70 **	12.35 ± 4.19	6.41 ± 4.27	23.28 ± 4.46 †
	SLP 75%	10.97 ± 2.70 **	-	23.32 ± 4.43 †	17.38 ± 4.58 **	34.25 ± 4.15 †
	BIOFIT-LP 60%	−12.35 ± 4.19	−23.32 ± 4.43 †	-	−5.94 ± 1.51 **	10.94 ± 2.41 **
	BIOFIT-LP 75%	−6.41 ± 4.27	−17.38 ± 4.58 **	5.94 ± 1.51 **	-	16.87 ± 2.92 †
	OFE-SLP	−23.28 ± 4.46 †	−34.25 ± 4.15 †	−10.94 ± 2.41 **	−16.87 ± 2.92 †	-
Tibialis Anterior	SLP 60%	-	−19.16 ± 3.49 †	−102.04 ± 14.40 †	−104.26 ± 16.56 †	−15.20 ± 11.15
	SLP 75%	19.16 ± 3.49 †	-	−82.89 ± 15.02 †	−85.11 ± 16.76 †	3.95 ± 11.70
	BIOFIT-LP 60%	102.04 ± 14.40 †	82.89 ± 15.02 †	-	−2.22 ± 9.75	86.84 ± 13.62 †
	BIOFIT-LP 75%	104.26 ± 16.56 †	85.11 ± 16.76 †	2.22 ± 9.75	-	89.06 ± 17.86 †
	OFE-SLP	15.20 ± 11.15	−3.95 ± 11.70	−86.84 ± 13.62 †	−89.06 ± 17.86 †	-
Gastrocn-emius	SLP 60%	-	−12.38 ± 2.44 †	17.11 ± 4.54 **	10.67 ± 4.70	10.86 ± 4.30
	SLP 75%	12.38 ± 2.44 †	-	29.48 ± 5.20 †	23.04 ± 5.32 **	23.24 ± 4.90 **
	BIOFIT-LP 60%	−17.11 ± 4.54 **	−29.48 ± 5.20 †	-	−6.44 ± 1.73 **	−6.25 ± 3.94
	BIOFIT-LP 75%	−10.67 ± 4.70	−23.04 ± 5.32 **	6.44 ± 1.73 **	-	0.19 ± 4.47
	OFE-SLP	−10.86 ± 4.30	−23.24 ± 4.90 **	6.25 ± 3.94	−0.19 ± 4.47	-

BIOFIT-LP = BIOFIT leg press; OFE-SLP = classic OFE; SLP = seated leg press ** = p value < 0.01; † = p value < 0.001.

4. Discussion

The main objective of this research was to analyze and compare the EMG activity of the muscle groups of the lower limbs produced during the execution of the leg press with a traditional SLP machine, and OFE-SLP and BIOFIT-LP machines. The results showed that in all three machines, the most activated muscle group was the vastus lateralis, followed by the rectus femoris and tibialis anterior. These results coincide with those found in a similar study, in which a high activation of the vastus lateralis, as well as the rectus femoris and tibialis anterior, was observed during the use of an SLP [22,27,40]. In light of these results, the performance of exercises involving simultaneous ankle and knee extension, regardless of the small biomechanical differences between exercises, involves the same major muscle groups [11]. This corroborates the findings of recent studies, in which muscle electrical activity was analyzed in exercises such as the SLP, the leg press at 45°, and the squat, finding that they all activate the same major muscle groups, despite their differences in execution, as they are all lower limb press exercises [1,11,14]. Furthermore, as all exercises are performed on stable, guided machines, the core musculature may be more involved, as they rely less on the stabilizing musculature, as compared to exercises with free weights [41] Therefore, it is possible to proceed to the comparison of the machines with each other.

In this regard, an outstanding result of the present investigation was that when comparing machines with the same percentage of intensity, more specifically the EMG activities measured in the SLP and the BIOFIT-LP, similar percentages of activation were found both at 60% and 75% of the 1RM in the main muscle groups involved, in the full stroke, and in the concentric and eccentric phases. No previous studies have analyzed EMG activity in OFE with a selector for the intensity of the exercise, and no studies were found that compared the same model of indoor and outdoor machines, with the present study being a pioneer in this aspect. However, in other studies comparing indoor machines designed to work the same muscle group, it has been found that at similar percentages of 1RM, the EMG activity of the main muscles involved is similar [19], confirming the results of the present study. In light of these results, it can be affirmed, along the same lines as previous studies, that the amplitude of the EMG was greatly dependent on intensity rather than on the execution technique for both traditional SLP and BIOFIT-LP machines [42].

However, a significantly greater activation was observed in the BIOFIT-LP machine as compared to the SLP machine for the tibialis anterior at both intensities in the full stroke and concentric phase, but not the eccentric one. In contrast, the gastrocnemius showed a greater activation in the SLP as compared to the BIOFIT-LP machine at both intensities and all phases, and the biceps femoris showed greater activation in the SLP at 75% 1RM during the eccentric phase. As observed in several studies, the position of the selector of the intensity of the exercise and ergonomic and biomechanical characteristics of the equipment with respect to the joint levers condition the participation of certain muscle groups [3,43,44]. In this sense, the findings for the SLP machine are in line with what was found in previous studies, where it was shown that in this machine, an increased activation of the biceps femoris and gastrocnemius was generated, especially in the push-off phase [4], due to the co-contraction of these muscle groups, together with the extensor musculature of the knee, to improve the anteroposterior stability of the joint in the angulations of greater instability [4]. Another possible explanation is that the projection of the intensity of the exercise in the SLP falls forward with respect to the base of support and the support of the feet, so that the co-activation of the biceps femoris and gastrocnemius is vital for knee stabilization [4,45]. In contrast, in the BIOFIT-LP machine, the intensity of the exercise projection falls further back with respect to the base of support and foot support, which could favor the activation of the tibialis anterior as a stabilizer of the ankle joint [46]. For all of the above, in general terms, it can be stated that with the new BIOFIT-LP design, an EMG similar to that achieved with the traditional fitness room machines, such as the SLP, could be achieved. Therefore, this new machine design could provide an alternative for effective strength training, albeit outdoors and free of charge, with the advantages that this could have for the health of the population [8]. However, the lack of previous studies analyzing this issue with OFE, and the fact that the studies that have analyzed EMG activity with traditional SLP have included fewer muscle groups than in the present investigation [19], makes it necessary to conduct future research that contrasts the results of the present investigation.

Another relevant result of the present investigation was that when comparing the EMG of the BIOFIT-LP and SLP, with respect to the OFE-SLP, the machines with an external selector for the intensity of the exercise showed significantly higher values in all muscle groups, regardless of the intensity of the exercise, with a few exceptions. Despite the popularity of OFE in recent years [8,24,25], no previous study was found that analyzed EMG activity on this type of machine, so that its efficacy is somewhat uncertain [47]. However, previous studies have pointed out that intensity selection conditions fiber recruitment, and therefore the recorded signal [48,49]. Considering that the classic OFE line only allows working with self-weight [24], the results of the present investigation could be due to the fact that the subjects' own weight did not represent an intensity close to their 60% or 75% of the 1RM during the use of the OFE-SLP. Therefore, in light of the results of the present investigation, the OFE-SLP machine may have a limited efficacy, given the EMG activity it generates [47]. Therefore, it would have little capacity to generate adaptations, especially in terms of hypertrophy in most of the population, as it is lacking the increased muscle recruitment that can potentially provide a more anabolic stimulus [6], being able to only provide a sufficient stimulus in populations with low fitness levels, such as older detrained populations [23].

However, when comparing the EMG of the OFE-SLP with respect to the other two machines, no differences were found in the tibialis anterior as compared to the SLP for the full stroke and eccentric phase; for the gastrocnemius as compared to the BIOFIT-LP for the full stroke, and the concentric and eccentric phases; and for the gastrocnemius as compared to the SLP at 60% 1RM in the eccentric phase. The discussion of this section is limited, as only one article was found that analyzed the activity of gastrocnemius and tibialis anterior at the same time during the use of guided machines for lower limb training [40], developed on classic fitness center machinery, without including any type of OFE in the analysis. In any case, the function of the gastrocnemius and tibialis anterior muscle groups in this type of exercise is stabilization [4,19], so the lack of differences in EMG between machines may be due to the fact that the execution of these exercises, when performed with guided machines, may not be greatly dependent on these stabilizers [4].

The second objective of this work was to determine the differences in the EMG of the muscle groups of the lower limbs at 60 and 75% of the 1RM on machines that allow selection of the intensity of the exercise (SLP and BIOFIT-LP). The results showed an increase in the EMG activity of all the muscle groups analyzed when performing at an intensity of 75% of the 1RM, compared to 60% of the 1RM, both for the full stroke and for the concentric and eccentric phases in the SLP. The results are in line with those found in previous studies, in which an increase in EMG was found with an increasing intensity [22]. This is because at higher intensities, the recruitment and firing frequency of the recruited motor units increases [48,49].

In contrast, the BIOFIT-LP machine showed significant differences with increasing intensity only in the vastus lateralis for the full stroke and concentric phase, in the biceps femoris for the full stroke and concentric and eccentric phases, and in the gastrocnemius for the eccentric phase. No differences were found in any of the EMG cases on this machine with an increasing intensity for the rectus femoris and tibialis anterior. The fact that differences were found in the posterior musculature in the eccentric phase could be due to the fact that this phase in this exercise is performed in favor of gravity, and the participants were asked to perform it at a controlled speed, so that in the face of an increase in the intensity, it is possible that there would be an increase in EMG activation responsible for slowing down the movement, i.e., posterior musculature [14,50]. In contrast, the concentric phase is performed against gravity, which might require a greater activation of the biceps femoris musculature as the main knee stabilizer [14]. Regarding the vastus lateralis and rectus femoris, the intensity could have been insufficient for the recruitment of more fibers by the rectus femoris, showing differences only in the vastus lateralis. Previous studies have indicated that at high intensity (90% of the 1RM), the vastus lateralis presents greater activations than the rectus femoris [22]. For this same reason, the signal recorded in the tibialis anterior did not present significant differences even when increasing the intensity, indicating the lack of a different stimulus in terms of joint stabilization [46]. Based on the results of the present investigation, in general terms, an increase in the intensity could imply an increase in the recruitment of fibers of the main musculature involved, and therefore, of the adaptations that will be achieved by training with such machinery [22].

The main practical application of this research is that a new line of OFE that has a series of adaptations to be able to adjust the intensity of the exercise achieves similar EMG activations to conventional gym machinery. This information is valuable for both users and trainers, as it demonstrates that this line of machines could be just as effective as conventional machinery to train lower limbs in trained young men. Most of these machines are located in free, open-access facilities, providing an opportunity for the population to do strength work in an economical way and outdoors. The present research also shows that when this type of OFE does not have a system to adjust the intensity of the exercise, a lower EMG intensity is generated, so that the use of OFE that only allows training with one's own body weight could be less effective for strength training than machines that allow an adjustment of the intensity of the exercise in trained young men. Therefore, it would be advisable for public agencies that are investing in the installation of this OFE in public spaces to consider this in order to choose models that have an adjustment of the intensity of the exercise, thus maximizing the population that will be able to train effectively in these facilities. Regarding possible future lines of research, it would be advisable to analyze other prototypes of OFE to study the effect of different design changes on the EMG presented during their use in the different muscle groups. On the other hand, in the current study, the effect of fatigue was mitigated by allowing a minute of rest between each repetition [51]. Even so, future studies should use one of the various methods of analysis of neuromuscular fatigue [52] as well as zero-crossing of EMG signal [53]. In addition, it would be recommended to replicate the study in a female population or in other age populations to analyze whether the results are replicable in other populations.

Despite the novelty of the present study with respect to previous research on the analysis of the effectiveness of OFE, the present investigation is not without limitations. First, the EMG analysis of the gluteus as a hip extensor was not included due to the location of the electrodes being incompatible with the execution of the exercise on the machines, which is performed from sedentation (SLP and BIOFIT-LP). As a second limitation, the three machines exhibited small variations in the biomechanics of execution, which are conditioned by the mechanism of each of the structures, which makes the comparison between the exercises not exact. As a third limitation, the present study did not include the maximum voluntary contraction test as a variable in the EMG analysis, nor did it record the EMG during the 1RM test, variables that could have helped to differentiate the percentage of activation of each muscle group with respect to its maximum for each of the exercises involved. However, the EMG activation tests on the different machines were performed on the same day, following a randomized order, to eliminate the effect of daily variation on this variable. As a final limitation, although an in situ goniometric analysis was used to differentiate the concentric and eccentric phase, the present study did not include a goniometric analysis of the ankle, knee, and hip joints in the data analysis, which could have helped to discern the origin of the difference in EMG activation between some of the machines.

5. Conclusions

It was observed that the EMG activity presented by the current OFE, designed to strengthen the lower limb musculature (OFE-SLP), is significantly lower than that of the SLP, a machine commonly used in strength training in fitness centers by trained young men. On the other hand, it was found that when some changes in the machine configuration are made to the OFE, and an external intensity of the exercise is included, as performed on the BIOFIT-LP machine, activations similar to those of the SLP are achieved in trained young men. Moreover, in general terms, a higher intensity implied a higher activation of the involved musculature in trained young men. In light of these results, OFE could be considered an alternative to traditional machines for strength work in trained young men.

6. Patents

This research includes data on a new machine for the outdoor fitness equipment line, which is in the process of patenting by the Spanish Patent and Trademark Office (code: ES2975897), as well as of an external intensity of the exercise selection system for this machine (code: ES2975886).

Supplementary Materials: The following supporting information can be downloaded at: https://www.mdpi.com/article/10.3390/app14167390/s1, Table S1: Research model.

Author Contributions: Conceptualization, T.A.-L., P.J.M.-P., N.G.-G., A.E.-G. and R.V.-C.; methodology, T.A.-L., P.J.M.-P., N.G.-G., A.E.-G. and R.V.-C.; formal analysis, T.A.-L. and N.G.-G.; investigation, T.A.-L. and A.E.-G.; resources, F.E.-R. and R.V.-C.; data curation, T.A.-L., N.G.-G. and A.E.-G.; writing—original draft preparation, T.A.-L. and R.V.-C.; writing—review and editing, T.A.-L., P.J.M.-P., J.A.A., N.G.-G., A.E.-G., F.E.-R. and R.V.-C.; visualization, J.A.A.; supervision, R.V.-C.; project administration, P.J.M.-P. and N.G.-G.; funding acquisition, P.J.M.-P., J.A.A. and N.G.-G. All authors have read and agreed to the published version of the manuscript.

Funding: This project was funded by the Ministry of Science, Innovation and Universities of the Government of Spain in the call Retos-Colaboración 2017, under the title RTC-2017-6145-1, 2017. T.A.-L.'s participation in the present research is the result of a 2020–2021 Research Staff Training grant awarded by the UCAM Universidad Católica de Murcia.

Institutional Review Board Statement: This study was conducted in accordance with the Declaration of Helsinki, and approved by the Institutional Ethics Committee of the Universidad Católica de Murcia (Ethical Application Ref: CE111908; 29 November 2019).

Informed Consent Statement: Informed consent was obtained from all subjects involved in the study.

Data Availability Statement: The data presented in this study are available on request from the corresponding author due to privacy.

Acknowledgments: We would like to thank the volunteers for their participation in the study. In addition, we would also thank the company Entorno Urbano S.L.U., Copele S.L.U., the Centro Tecnológico del Metal (CTM) of the Region of Murcia for providing us with the outdoor fitness equipment and managing transport and assembly of the equipment. T.A.-L. would like to thank the UCAM Universidad Católica de Murcia for its support through the FPI grant for the development of this research toward his doctoral thesis. This paper is part of the doctoral thesis of T.A.-L.

Conflicts of Interest: The authors declare no conflicts of interest.

References

1. Armstrong, R.; Baltzopoulos, V.; Langan-Evans, C.; Clark, D.; Jarvis, J.; Stewart, C.; O'Brien, T. An Investigation of Movement Dynamics and Muscle Activity during Traditional and Accentuated-Eccentric Squatting. *PLoS ONE* **2022**, *17*, e0276096. [CrossRef]
2. Fragala, M.S.; Cadore, E.L.; Dorgo, S.; Izquierdo, M.; Kraemer, W.J.; Peterson, M.D.; Ryan, E.D. Resistance Training for Older Adults: Position Statement From the National Strength and Conditioning Association. *J. Strength Cond. Res.* **2019**, *33*, 2019–2052. [CrossRef]
3. Martín-Fuentes, I.; Oliva-Lozano, J.M.; Muyor, J.M. Electromyographic Activity in Deadlift Exercise and Its Variants. A Systematic Review. *PLoS ONE* **2020**, *15*, e0229507. [CrossRef] [PubMed]
4. Escamilla, R.F.; Fleisig, G.S.; Zheng, N.; Lander, J.E.; Barrentine, S.W.; Andrews, J.R.; Bergemann, B.W.; Moorman, C.T. Effects of Technique Variations on Knee Biomechanics during the Squat and Leg Press. *Med. Sci. Sports Exerc.* **2001**, *33*, 1552–1566. [CrossRef] [PubMed]
5. Sarto, F.; Franchi, M.V.; Rigon, P.A.; Grigoletto, D.; Zoffoli, L.; Zanuso, S.; Narici, M.V. Muscle Activation during Leg-Press Exercise with or without Eccentric Overload. *Eur. J. Appl. Physiol.* **2020**, *120*, 1651–1656. [CrossRef] [PubMed]
6. Schwanbeck, S.R.; Cornish, S.M.; Barss, T.; Chilibeck, P.D. Effects of Training with Free Weights Versus Machines on Muscle Mass, Strength, Free Testosterone, and Free Cortisol Levels. *J. Strength Cond. Res.* **2020**, *34*, 1851–1859. [CrossRef]
7. Abelleira-Lamela, T.; Vaquero-Cristóbal, R.; González-Gálvez, N.; Esparza-Ros, F.; Espeso-García, A.; Marcos-Pardo, P.J. Sagittal Spine Disposition and Pelvic Tilt during Outdoor Fitness Equipment Use and Their Associations with Kinanthropometry Proportions in Middle-Aged and Older Adults. *PeerJ* **2021**, *9*, e12657. [CrossRef]
8. Chow, H.; Wu, D.-R. Outdoor Fitness Equipment Usage Behaviors in Natural Settings. *Int. J. Environ. Res. Public Health* **2019**, *16*, 391. [CrossRef]
9. Marcos-Pardo, P.J.; Espeso-García, A.; Vaquero-Cristóbal, R.; Abelleira-Lamela, T.; González-Gálvez, N. The Effect of Resistance Training with Outdoor Fitness Equipment on the Body Composition, Physical Fitness, and Physical Health of Middle-Aged and Older Adults: A Randomized Controlled Trial. *Healthcare* **2024**, *12*, 726. [CrossRef]
10. Handsfield, G.G.; Meyer, C.H.; Hart, J.M.; Abel, M.F.; Blemker, S.S. Relationships of 35 Lower Limb Muscles to Height and Body Mass Quantified Using MRI. *J. Biomech.* **2014**, *47*, 631–638. [CrossRef]
11. Da Silva, E.M.; Brentano, M.A.; Cadore, E.L.; De Almeida, A.P.V.; Kruel, L.F.M. Analysis of Muscle Activation During Different Leg Press Exercises at Submaximum Effort Levels. *J. Strength Cond. Res.* **2008**, *22*, 1059–1065. [CrossRef] [PubMed]
12. van den Tillaar, R.; Kristiansen, E.L.; Larsen, S. Is the Occurrence of the Sticking Region in Maximum Smith Machine Squats the Result of Diminishing Potentiation and Co-Contraction of the Prime Movers among Recreationally Resistance Trained Males? *Int. J. Environ. Res. Public Health* **2021**, *18*, 1366. [CrossRef] [PubMed]
13. Alizadeh, S.; Rayner, M.; Mamdouh, M.; Mahmoud, I.; Behm, D.G.; John', S.; Labrador, C. Push-Ups vs. Bench Press Differences in Repetitions and Muscle Activation between Sexes. *J. Sports Sci. Med.* **2020**, *19*, 289–297.
14. van den Tillaar, R.; Andersen, V.; Saeterbakken, A.H. Comparison of Muscle Activation and Kinematics during Free-Weight Back Squats with Different Loads. *PLoS ONE* **2019**, *14*, e0217044. [CrossRef] [PubMed]
15. Miyajima, T.; Ishida, K.; Sato, M.; Yamagata, Z. Pilot Study to Test the Safety of an Exercise Machine on Healthy Adult Females. *Jpn. J. Nurs. Sci.* **2010**, *7*, 37–46. [CrossRef] [PubMed]

16. Krause Neto, W.; Vieira, L.; Gama, E.F. Barbell Hip Thrust, Muscular Activation and Performance: A Systematic Review. *J. Sports Sci. Med.* **2019**, *18*, 198–206.
17. Llurda-Almuzara, L.; Labata-Lezaun, N.; López-de-Celis, C.; Aiguadé-Aiguadé, R.; Romaní-Sánchez, S.; Rodríguez-Sanz, J.; Fernández-de-las-Peñas, C.; Pérez-Bellmunt, A. Biceps Femoris Activation during Hamstring Strength Exercises: A Systematic Review. *Int. J. Environ. Res. Public Health* **2021**, *18*, 8733. [CrossRef] [PubMed]
18. Zambrano, H.; Torres, X.; Coleman, M.; Franchi, M.V.; Fisher, J.P.; Oberlin, D.; Van Hooren, B.; Swinton, P.A.; Schoenfeld, B.J. Myoelectric Activity during Electromagnetic Resistance Alone and in Combination with Variable Resistance or Eccentric Overload. *Sci. Rep.* **2023**, *13*, 8212. [CrossRef]
19. Martín-Fuentes, I.; Oliva-Lozano, J.M.; Muyor, J.M. Evaluation of the Lower Limb Muscles' Electromyographic Activity during the Leg Press Exercise and Its Variants: A Systematic Review. *Int. J. Environ. Res. Public Health* **2020**, *17*, 4626. [CrossRef]
20. Trebs, A.A.; Brandenburg, J.P.; Pitney, W.A. An Electromyography Analysis of 3 Muscles Surrounding the Shoulder Joint during the Performance of a Chest Press Exercise at Several Angles. *J. Strength Cond. Res.* **2010**, *24*, 1925–1930. [CrossRef]
21. Lee, H.; Jung, M.; Lee, K.-K.; Lee, S. A 3D Human-Machine Integrated Design and Analysis Framework for Squat Exercises with a Smith Machine. *Sensors* **2017**, *17*, 299. [CrossRef] [PubMed]
22. Gonzalez, A.M.; Ghigiarelli, J.J.; Sell, K.M.; Shone, E.W.; Kelly, C.F.; Mangine, G.T. Muscle Activation during Resistance Exercise at 70% and 90% 1-repetition Maximum in Resistance-trained Men. *Muscle Nerve* **2017**, *56*, 505–509. [CrossRef] [PubMed]
23. Kim, D.-I.; Lee, D.H.; Hong, S.; Jo, S.; Won, Y.; Jeon, J.Y. Six Weeks of Combined Aerobic and Resistance Exercise Using Outdoor Exercise Machines Improves Fitness, Insulin Resistance, and Chemerin in the Korean Elderly: A Pilot Randomized Controlled Trial. *Arch. Gerontol. Geriatr.* **2018**, *75*, 59–64. [CrossRef]
24. Liu, Y.-C.; Yang, W.-W.; Fang, I.-Y.; Pan, H.L.-L.; Chen, W.-H.; Liu, C. Training Program with Outdoor Fitness Equipment in Parks Offers No Substantial Benefits for Functional Fitness in Active Seniors: A Randomized Controlled Trial. *J. Aging Phys. Act.* **2020**, *28*, 828–835. [CrossRef] [PubMed]
25. Chow, H.W.; Chang, K.T.; Fang, I.Y. Evaluation of the Effectiveness of Outdoor Fitness Equipment Intervention in Achieving Fitness Goals for Seniors. *Int. J. Environ. Res. Public Health* **2021**, *18*, 12508. [CrossRef]
26. Bhalerao, S.; Kadam, P. Sample Size Calculation. *Int. J. Ayurveda Res.* **2010**, *1*, 55. [CrossRef]
27. Schoenfeld, B.J.; Contreras, B.; Willardson, J.M.; Fontana, F.; Tiryaki-Sonmez, G. Muscle Activation during Low- versus High-Load Resistance Training in Well-Trained Men. *Eur. J. Appl. Physiol.* **2014**, *114*, 2491–2497. [CrossRef]
28. Esparza-Ros, F.; Vaquero-Cristóbal, R.; Marfell-Jones, M.J. *International Standards for Anthropometric Assessment -Full Profile-*; International Society for Advancement in Kinanthropometry: Murcia, Spain, 2019.
29. Miller, T.A. *NSCA's Guide to Tests and Assessments*; Human Kinetics: Champaign, IL, USA, 2012.
30. López-Vivancos, A.; González-Gálvez, N.; Orquín-Castrillón, F.J.; Vale, R.G.d.S.; Marcos-Pardo, P.J. Electromyographic Activity of the Pectoralis Major Muscle during Traditional Bench Press and Other Variants of Pectoral Exercises: A Systematic Review and Meta-Analysis. *Appl. Sci.* **2023**, *13*, 5203. [CrossRef]
31. Stegeman, D.F.; Hermens, H.J. *Standards for Surface Electromyography: The European Project "Surface EMG for Non-Invasive Assessment of Muscles (SENIAM)"*; Roessingh Research and Development: Enschede, The Netherlands, 2007.
32. Snarr, R.L.; Esco, M.R. Electromyographic Comparison of Traditional and Suspension Push-Ups. *J. Hum. Kinet.* **2013**, *39*, 75–83. [CrossRef]
33. Carius, D.; Kugler, P.; Kuhwald, H.-M.; Wollny, R. Absolute and Relative Intrasession Reliability of Surface EMG Variables for Voluntary Precise Forearm Movements. *J. Electromyogr. Kinesiol.* **2015**, *25*, 860–869. [CrossRef]
34. Mitchell, L.J.; Argus, C.K.; Taylor, K.-L.; Sheppard, J.M.; Chapman, D.W. The Effect of Initial Knee Angle on Concentric-Only Squat Jump Performance. *Res. Q. Exerc. Sport* **2017**, *88*, 184–192. [CrossRef] [PubMed]
35. Rossi, F.E.; Schoenfeld, B.J.; Ocetnik, S.; Young, J.; Vigotsky, A.; Contreras, B.; Krieger, J.W.; Miller, M.G.; Cholewa, J. Strength, Body Composition, and Functional Outcomes in the Squat versus Leg Press Exercises. *J. Sports Med. Phys. Fit.* **2018**, *58*, 263–270. [CrossRef]
36. Andersen, V.; Fimland, M.S.; Mo, D.A.; Iversen, V.M.; Vederhus, T.; Rockland Hellebø, L.R.; Nordaune, K.I.; Saeterbakken, A.H. Electromyographic Comparison of Barbell Deadlift, Hex Bar Deadlift, and Hip Thrust Exercises: A Cross-over Study. *J. Strength Cond. Res.* **2018**, *32*, 587–593. [CrossRef]
37. Picerno, P.; Iannetta, D.; Comotto, S.; Donati, M.; Pecoraro, F.; Zok, M.; Tollis, G.; Figura, M.; Varalda, C.; Di Muzio, D.; et al. 1RM Prediction: A Novel Methodology Based on the Force–Velocity and Load–Velocity Relationships. *Eur. J. Appl. Physiol.* **2016**, *116*, 2035–2043. [CrossRef] [PubMed]
38. Padulo, J.; Annino, G.; Tihanyi, J.; Calcagno, G.; Vando, S.; Smith, L.; Vernillo, G.; La Torre, A.; D'ottavio, S. Uphill Racewalking at Iso-Efficiency Speed. *J. Strength Cond. Res.* **2013**, *27*, 1964–1973. [CrossRef] [PubMed]
39. dos Santos Albarello, J.C.; Cabral, H.V.; Leitão, B.F.M.; Halmenschlager, G.H.; Lulic-Kuryllo, T.; da Matta, T.T. Non-Uniform Excitation of Pectoralis Major Induced by Changes in Bench Press Inclination Leads to Uneven Variations in the Cross-Sectional Area Measured by Panoramic Ultrasonography. *J. Electromyogr. Kinesiol.* **2022**, *67*, 102722. [CrossRef] [PubMed]
40. Hahn, D. Lower Extremity Extension Force and Electromyography Properties as a Function of Knee Angle and Their Relation to Joint Torques: Implications for Strength Diagnostics. *J. Strength Cond. Res.* **2011**, *25*, 1622–1631. [CrossRef]
41. Cotterman, M.L.; Darby, L.A.; Skelly, W.A. Comparison of Muscle Force Production Using the Smith Machine and Free Weights for Bench Press and Squat Exercises. *J. Strength Cond. Res.* **2005**, *19*, 169. [CrossRef]

42. Bobet, J.; Norman, R.W. Use of the Average Electromyogram in Design Evaluation Investigation of a Whole-Body Task. *Ergonomics* **1982**, *25*, 1155–1163. [CrossRef]
43. Camara, K.D.; Coburn, J.W.; Dunnick, D.D.; Brown, L.E.; Galpin, A.J.; Costa, P.B. An Examination of Muscle Activation and Power Characteristics While Performing the Deadlift Exercise with Straight and Hexagonal Barbells. *J. Strength Cond. Res.* **2016**, *30*, 1183–1188. [CrossRef]
44. Jo, E.; Valenzuela, K.A.; Leyva, W.; Rivera, J.; Tomlinson, K.; Zeitz, E. Electromyographic Examination of Hip and Knee Extension Hex Bar Exercises Varied by Starting Knee and Torso Angles. *Int. J. Exerc. Sci.* **2022**, *15*, 541–551. [PubMed]
45. Kristiansen, E.; Larsen, S.; Haugen, M.E.; Helms, E.; van den Tillaar, R. A Biomechanical Comparison of the Safety-Bar, High-Bar and Low-Bar Squat around the Sticking Region among Recreationally Resistance-Trained Men and Women. *Int. J. Environ. Res. Public Health* **2021**, *18*, 8351. [CrossRef] [PubMed]
46. Uzun, B.; Taylan, O.; Gültekin, B.; Havıtçıoğlu, H. Dynamic Measurements of Musculus Tibialis Anterior Ligaments with Different Angles. *J. Biomech.* **2011**, *44*, 2. [CrossRef]
47. Mehr, K. Surface Electromyography in Orthodontics—A Literature Review. *Med. Sci. Monit.* **2013**, *19*, 416–423. [CrossRef] [PubMed]
48. Cormie, P.; McGuigan, M.R.; Newton, R.U. Developing Maximal Neuromuscular Power Part 1—Biological Basis of Maximal Power Production. *Sports Med.* **2011**, *41*, 17–38. [CrossRef]
49. Yavuz, H.U.; Erdağ, D.; Amca, A.M.; Aritan, S. Kinematic and EMG Activities during Front and Back Squat Variations in Maximum Loads. *J. Sports Sci.* **2015**, *33*, 1058–1066. [CrossRef]
50. Oshikawa, T.; Morimoto, Y.; Kaneoka, K. Lumbar Lordosis Angle and Trunk and Lower-Limb Electromyographic Activity Comparison in Hip Neutral Position and External Rotation during Back Squats. *J. Phys. Ther. Sci.* **2018**, *30*, 434–438. [CrossRef] [PubMed]
51. Von Werder, S.C.F.A.; Kleiber, T.; Disselhorst-Klug, C. A Method for a Categorized and Probabilistic Analysis of the Surface Electromyogram in Dynamic Contractions. *Front. Physiol.* **2015**, *6*, 30. [CrossRef]
52. Vøllestad, N.K. Measurement of Human Muscle Fatigue. *J. Neurosci. Methods* **1997**, *74*, 219–227. [CrossRef]
53. Kim, H.; Lee, J.; Kim, J. Electromyography-Signal-Based Muscle Fatigue Assessment for Knee Rehabilitation Monitoring Systems. *Biomed. Eng. Lett.* **2018**, *8*, 345–353. [CrossRef]

MDPI AG
Grosspeteranlage 5
4052 Basel
Switzerland
Tel.: +41 61 683 77 34

Applied Sciences Editorial Office
E-mail: applsci@mdpi.com
www.mdpi.com/journal/applsci

www.ingramcontent.com/pod-product-compliance
Lightning Source LLC
Chambersburg PA
CBHW062103220926
51850CB00010B/2877
* 9 7 8 3 7 2 5 8 6 7 5 6 1 *